AF832651

Wireless Networks

Series Editor

Xuemin Sherman Shen, University of Waterloo, Waterloo, Canada

The purpose of Springer's Wireless Networks book series is to establish the state of the art and set the course for future research and development in wireless communication networks. The scope of this series includes not only all aspects of wireless networks (including cellular networks, WiFi, sensor networks, and vehicular networks), but related areas such as cloud computing and big data. The series serves as a central source of references for wireless networks research and development. It aims to publish thorough and cohesive overviews on specific topics in wireless networks, as well as works that are larger in scope than survey articles and that contain more detailed background information. The series also provides coverage of advanced and timely topics worthy of monographs, contributed volumes, textbooks and handbooks.

Weiting Zhang · Dong Yang · Shuai Gao ·
Hongke Zhang · Xuemin Shen

Intelligent Resource Scheduling in End-Edge-Cloud Networks

Springer

Weiting Zhang
Beijing Jiaotong University
Beijing, China

Dong Yang
Beijing Jiaotong University
Beijing, China

Shuai Gao
Beijing Jiaotong University
Beijing, China

Hongke Zhang
Beijing Jiaotong University
Beijing, China

Xuemin Shen
Department of Electrical and Computer Engineering
University of Waterloo
Waterloo, ON, Canada

ISSN 2366-1186 ISSN 2366-1445 (electronic)
Wireless Networks
ISBN 978-3-032-07666-3 ISBN 978-3-032-07667-0 (eBook)
https://doi.org/10.1007/978-3-032-07667-0

© The Editor(s) (if applicable) and The Author(s), under exclusive license to Springer Nature Switzerland AG 2026

This work is subject to copyright. All rights are solely and exclusively licensed by the Publisher, whether the whole or part of the material is concerned, specifically the rights of translation, reprinting, reuse of illustrations, recitation, broadcasting, reproduction on microfilms or in any other physical way, and transmission or information storage and retrieval, electronic adaptation, computer software, or by similar or dissimilar methodology now known or hereafter developed.
The use of general descriptive names, registered names, trademarks, service marks, etc. in this publication does not imply, even in the absence of a specific statement, that such names are exempt from the relevant protective laws and regulations and therefore free for general use.
The publisher, the authors and the editors are safe to assume that the advice and information in this book are believed to be true and accurate at the date of publication. Neither the publisher nor the authors or the editors give a warranty, expressed or implied, with respect to the material contained herein or for any errors or omissions that may have been made. The publisher remains neutral with regard to jurisdictional claims in published maps and institutional affiliations.

This Springer imprint is published by the registered company Springer Nature Switzerland AG
The registered company address is: Gewerbestrasse 11, 6330 Cham, Switzerland

If disposing of this product, please recycle the paper.

Preface

The wide-spread deployment of intelligent applications is pushing the boundaries of traditional cloud computing. From real-time control in industry Internet of things (IIoT) to immersive experience in metaverse, these revolutionary applications expose critical limitations of traditional cloud-centric computing architectures, such as excessive end-to-end latency and severe bandwidth bottlenecks. Against this background, the end-edge-cloud collaborative architecture has emerged, which integrates the powerful computing resources of cloud servers, the low-latency processing capabilities of edge nodes, and the real-time sensing capabilities of end devices, ultimately forming a multi-level and distributed intelligent computing system. However, this integration requires not only functional aggregation, but also intelligent coordination of multi-dimensional resources in dynamic heterogeneous networks.

Specifically, the complexity of end-edge-cloud collaborative mainly manifests in two dimensions: (1) computation and (2) communication. In the computation dimension, how to achieve efficient collaborative computing has become a central focus for both academia and industry, especially for emerging artificial intelligence (AI)-based computing scenarios, such as federated learning (FL) training and deep neural network (DNN) inference. In such scenarios, the optimal allocation of heterogeneous computing resources and dynamic migration of cross-domain computing tasks constitute a highly complex optimization problem, which is difficult to be solved via traditional optimization methods. In the communication dimension, how to achieve efficient utilization of network resources while ensuring diversified requirements for various computing tasks becomes another key challenge. For example, industrial inspection and control requires deterministic transmission latency and near-zero packet loss, and real-time video analysis requires high bandwidth and stable transmission rate. Moreover, the end-edge-cloud collaborative architecture covers cellular networks, WiFi, and wired connections, which requires flexible resource scheduling to adapt to dynamic network conditions while satisfying strict quality of service (QoS) requirements for intelligent applications.

To address these challenges, this book proposes new network architectures and multi-dimensional resource scheduling schemes that integrate communication and computation to promote efficient FL model training and real-time DNN inference, thereby effectively supporting the full life cycle of various intelligent applications. This book contains six chapters:

In Chap. 1, we give an introduction to end-edge-cloud networks, including application scenarios, characteristics, resource scheduling technologies, and challenges faced in the communication and computation dimensions, and then the resource management in end-edge-cloud networks is introduced from different dimensions of resources.

In Chap. 2, we design $(\mathrm{Com})^2$Net, an integrated communication and computation network architecture that schedules computing task across space-air-ground, end-edge-cloud, and multi-data center domains, where a quantum genetic-based scheduling algorithm (QGA) is proposed to optimize inter-domain traffic scheduling and intra-domain resource allocation under QoS and quality of experience (QoE) constraints.

In Chap. 3, we develop a three-layer FL architecture for distributed IIoT networks, introducing the reinforcement on federated (RoF) scheme with multi-agent deep reinforcement learning (DRL) to jointly optimize device selection and resource allocation while minimizing FL loss and satisfying energy and latency requirements, achieving superior performance over existing benchmarks.

In Chap. 4, we propose DetFed, a deterministic FL framework that integrates time-sensitive networks (TSN) with cyclic queuing mechanisms, and presents a DRL-based dynamic scheduling algorithm that simultaneously maximizes the scheduling success ratio of parameter flows and learning accuracy of the FL model with strict deterministic transmission requirements.

In Chap. 5, we design an end-edge-cloud collaborative architecture to support efficient DNN inference. Moreover, we propose a DRL-based resource management solution to efficiently allocate available spectrum, cache, and computing resources, thereby improving the average inference accuracy.

In Chap. 6, we conclude the book by systematically summarizing the presented architectures and algorithms for end-edge-cloud resource scheduling, while discussing future research directions in network orchestration and learning-based resource management.

We would like to thank Prof. Haixia Peng from Xi'an Jiaotong University, Prof. Wen Wu from Peng Cheng Laboratory, and Dr. Chengrui Zhang and Tong Sun from Beijing Jiaotong University for their contributions to the book. We would also like to thank all the group members for their valuable suggestions and comments. Special thanks go to the senior editor Mary E. James and Pradheepa Vijay from Springer Nature for their help throughout the publication process.

Beijing, China — Weiting Zhang
Beijing, China — Dong Yang
Beijing, China — Shuai Gao
Beijing, China — Hongke Zhang
Waterloo, ON, Canada — Xuemin Shen

Acknowledgements This work was supported by the National Natural Science Foundation of China under Grants 62394320 and 62425104.

Competing Interests The authors have no competing interests to declare that are relevant to the content of this manuscript.

Contents

Acronyms

3GPP	Third Generation Partnership Project
ACRNN	Attention-Enhanced Convolutional Recurrent Neural Network
AI	Artificial Intelligence
APLs	Automated Production Lines
BS	Base Station
CCaS	Cloud Caching Server
CNN	Convolutional Neural Network
Conv	Convolutional
CPU	Central Processing Unit
CQF	Cyclic Queuing and Forwarding
CT	Centralized Training
D2D	Device-to-Device
D3QN	Dueling Double Deep Q-network
DCPU	DNN Caching Pool Updating
DDPG	Deep Deterministic Policy Gradient
DDQN	Double Deep Q-Network
DNN	Deep Neural Network
DQN	Deep Q-Network
DRL	Deep Reinforcement Learning
ECaS	Edge Caching Server
ECoS	Edge Computing Server
EECS	End-Edge-Cloud Synergy
eMBB	Enhanced Mobile Broadband
FC	Fully-Connected
FCNN	Four-Layer Fully-Connected Neural Network
FedAvg	Federated Averaging
FFT	Fast Fourier Transform
FL	Federated Learning
FPGA	Field Programmable Gate Array
GA	Genetic Algorithm
GCL	Gate Control Lists

GPU	Graphics Processing Unit
HSR	High-Speed Railways
IGWs	Industrial Gateways
IIoT	Industrial Internet of Things
INC	In-Network Computing
IoE	Intelligence of Everything
JDSRA	Joint Device Selection and Resource Allocation
LSTM	Long Short-Term Memory
MADDPG	Multi-Agent Deep Deterministic Policy Gradient
MASAC	Multi-Agent Soft Actor Critic
MDP	Markov Decision Process
MEC	Multi-Access Edge Computing
MILP	Mixed-Integer Linear Programming
MMTC	Massive Machine Type Communications
NFV	Network Function Virtualization
NIC	Network Interface Card
Non-IID	Non-Independent and Identically Distributed
ONC	Out-Network Computing
Pi 4B	Raspberry Pi 4 Model B
POMDP	Partially-Observable Markov Decision Process
PTP	Precision Time Protocol
QGAs	Quantum Genetic Algorithms
QoE	Quality of Experience
QoS	Quality of Service
RAN	Radio Access Networks
RL	Reinforcement Learning
RoF	Reinforcement on Federated
RPC	Random Probabilistic Configuration
RS	Random-based Scheme
SARoF	Single-Agent RoF
SRP	Stream Reservation Protocol
TADP	Task Assignment and DNN Placement
TAS	Time-Aware Shaper
TD3	Twin Delayed Deep Deterministic Policy Gradient
TSN	Time-Sensitive Networks
UAV	Unmanned Aerial Vehicle
URLLC	Ultra-Reliable Low-Latency Communications
USR	Ultrahigh-Speed Railway
VR	Virtual Reality

Chapter 1
Introduction

1.1 Overview of End-Edge-Cloud Networks

The end-edge-cloud network represents an emerging paradigm in the field of distributed computing, characterized by the seamless integration of end devices, edge nodes, and centralized cloud resources into a unified hierarchical architecture [1]. The architecture fully leverages the complementary advantages of each layer, with end devices providing ubiquitous data generation and preliminary processing capabilities, edge nodes offering task offloading support near the data source and enabling low-latency response, and the cloud delivering large-scale storage and powerful computing capabilities. Through the implementation of dynamic collaboration and resource sharing across heterogeneous layers, end-edge-cloud networks effectively overcome the limitations of traditional cloud-centric computing architectures, such as high latency, network congestion, and data privacy risks. Furthermore, in the context of exponentially growing data volumes, the end-edge-cloud network establishes a solid foundation for supporting intelligent, adaptive, and scalable services. In this section, we will first introduce the application scenarios and characteristics of end-edge-cloud networks. Then, we outline the distinctive features of federated learning (FL) as a representative approach to distributed computing and summarize resource scheduling technologies available for end-edge-cloud networks (Fig. 1.1).

1.1.1 End-Edge-Cloud Network Applications

End-edge-cloud networks have become an essential paradigm for supporting a wide range of applications by enabling low-latency processing, efficient resource utilization, and scalable computing capabilities. These networks are particularly well-suited to fields where real-time response and intelligent decision-making are critical, including industrial automation, intelligent transportation, smart city infrastructures,

© The Author(s), under exclusive license to Springer Nature Switzerland AG 2026

W. Zhang et al., *Intelligent Resource Scheduling in End-Edge-Cloud Networks*,
Wireless Networks, https://doi.org/10.1007/978-3-032-07667-0_1

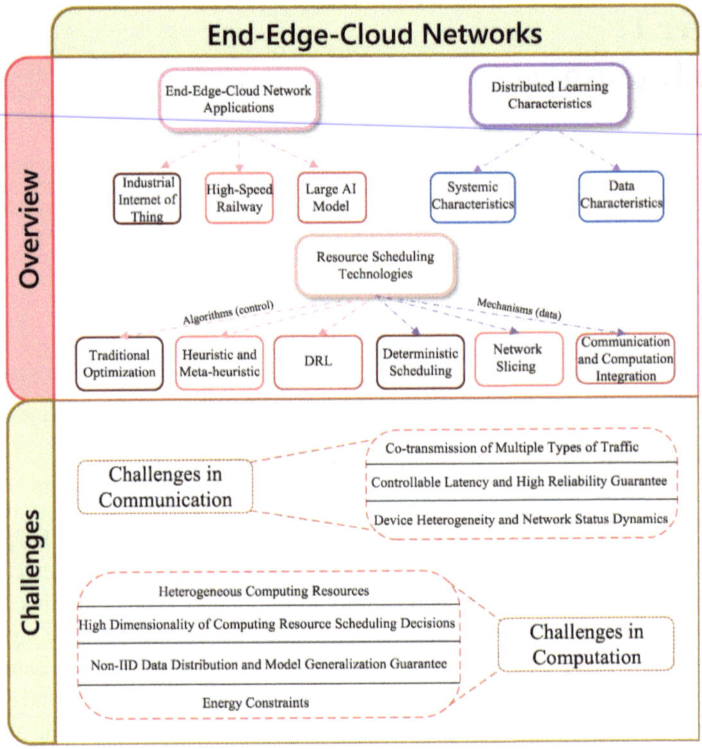

Fig. 1.1 An illustration of end-edge-cloud networks

healthcare systems, and others. This part focuses on three prominent application scenarios: industrial Internet of things (IIoT), high-speed railways, and large artificial intelligence (AI) models.

IIoT: End-edge-cloud collaborative IIoT architecture combines the advantages of local computing, edge computing, and cloud computing, which can help effectively address the diversified service requirements in IIoT [2]. The end layer consists of various kinds of IIoT devices in production lines, such as traditional machine tools, smart equipment, data collectors, sensors, and industrial robots. Every device continuously produces various computing tasks and is responsible for real-time data collection and initial processing, making them suitable for latency-sensitive tasks that require immediate responses. Every production line is equipped with a corresponding edge computing server that has certain computing resources and associated with a base station (BS) for data transmission to form the edge layer. Additionally, the edge layer includes numerous switches that form a switching network to facilitate data transmission among edge servers. The cloud layer offers substantial computing resources to address the computing tasks. Moreover, it provides big data analytics and predictive maintenance services for the entire IIoT system.

High-Speed Railway (HSR): The train-edge-cloud architecture addresses the computing challenges of HSR systems by distributing tasks across multiple layers, providing reliable and efficient intelligent services for HSR systems. Equipped with onboard sensors and cameras, the train layer handles real-time data collection and latency-sensitive tasks such as fault detection and autonomous control. Due to limited onboard resources, complex tasks need to be offloaded to the edge or cloud layers [3]. The edge layer is composed of edge servers deployed along the railway tracks to provide moderate computing resources and acts as intermediaries between the trains and the cloud. Using high-speed 5G wireless communication, trains can offload their tasks to the nearby edge servers. The edge layer is responsible for handling tasks that require real-time processing and moderate computing resources, such as intrusion detection and localized scheduling. The cloud layer offers extensive computing resources to process complex and resource-intensive tasks that cannot be handled by the edge layer, including global optimization of train schedules and historical data analysis for predictive maintenance and fault diagnosis.

Large AI Model: The end-edge-cloud architecture provides support for the construction of large AI models in the entire lifecycle through collaboration across multiple layers, including training, inference, and deployment phases. During the training phase, the cloud provides massive computing resources for the training of large AI models, edge nodes facilitate efficient FL using the decentralized computing resources, and the end devices contribute local data for personalization [4]. During the inference phase, the tasks are strategically distributed, with the cloud handling resource-intensive processing, the edge nodes performing optimized models for latency-sensitive applications, and end devices running quantized models for immediate interaction. During the deployment phase, the cloud performs centralized control and updates, the edge nodes conducts context-specific model optimization and compression, and end devices ensures seamless integration and consistent user experience.

In summary, end-edge-cloud collaborative networks provide efficient resource support for diversified service requirements via effective computation and communication integration in various application fields, which is becoming a critical technological foundation for driving digital transformation and intelligent upgrades.

1.1.2 Distributed Learning Characteristics

End-edge-cloud networks facilitate the deployment of distributed AI technologies, such as FL, which effectively leverage the computing and communication resources dispersed throughout the network while preserving user privacy and reducing communication overhead. Specifically, FL enables multiple geographically dispersed devices or nodes to collaboratively train a shared global model. Each participating node utilizes its local data to update the model and uploads only the model parameters (e.g., gradients or weights) to a central server for aggregation, while the raw data remains stored locally. The characteristics of FL are not only reflected in decentralization and privacy protection, but also encompass multiple elements, including systems, algorithms, data, and security. To further elaborate on the characteristics of FL, we classify them into systemic characteristics and data characteristics.

Systemic Characteristics: The systemic characteristics, including distributed computing, communication efficiency, and high elasticity and fault tolerance, pose challenges to the resource allocation of networks.

(1) *Distributed Computing*: FL exhibits a highly distributed collaborative computing characteristic, where heterogeneous nodes across end, edge, and cloud layers are leveraged to collaboratively participate in model training. Within this framework, data and computing tasks are geographically dispersed, forming a decentralized computing paradigm through local model updates and global parameter aggregation, which significantly enhances the scalability and flexibility of the system, while effectively utilizing multi-dimensional resources including computing, communication, and storage throughout the network. However, the participating nodes demonstrate significant heterogeneity in terms of computing capability, bandwidth, energy consumption, and storage capacity. Moreover, the availability of these nodes and the network conditions are highly dynamic. The above factors pose substantial challenges to the rational allocation of tasks, efficient utilization of resources, synchronization of training progress, and the overall balance of system performance.

(2) *Communication Efficiency*: FL requires dispersed nodes to periodically upload local model parameters or gradients to a central server for global aggregation, after which the updated global model is then disseminated back to these nodes. This process imposes stringent demands on communication efficiency, as the frequent exchange of parameters not only increases communication overhead but also introduces synchronization delays, which can negatively impact the convergence speed and accuracy of the global model, particularly in scenarios involving a large number of participating nodes, multiple training iterations, and models with high parameter complexity. Therefore, efficient model compression and sparsification algorithms are urgently required to reduce the data volume transmitted in each FL round. Mechanisms such as hierarchical aggregation and local synchronization can further alleviate the burden on network transmission. Furthermore, the design of robust communication resource scheduling algorithms is essential to ensure high reliability in parameter transmission.

(3) *High Elasticity and Fault Tolerance*: FL demonstrates high elasticity and fault tolerance to adapt to the frequent dynamics of node availability in the end-edge-cloud networks. Nodes may join or leave the training process temporarily due to factors such as battery constraints or user behavior, resulting in a constantly changing set of participating nodes. In addition, issues such as network interruptions and hardware failures can lead to the failure of certain nodes. FL needs to be capable of real-time perception and adaptation to the dynamic nature of

participating node sets. Through fault tolerance mechanisms, including asynchronous updates and participating node selection, FL enhances the robustness of the overall training process.

Data Characteristics: The data characteristics, including data heterogeneity and privacy protection, put strict forward to the entire process of FL.

(1) *Data Heterogeneity and Non-independent and Identically Distributed (Non-IID)*: During the FL process, the data held by participating nodes is often highly heterogeneous, with data distributions typically exhibiting pronounced Non-IID characteristics. Firstly, this heterogeneity is manifested in differences in feature distributions, as data collected by different devices may experience distributional shifts in the feature space [5]. Secondly, it is also reflected in label imbalance, since some devices might only gather data from specific classes. Thirdly, there is often a significant disparity in data quantity among devices [6]. In response to these cases, various innovative mechanisms such as heterogeneity-aware aggregation, adaptive weight allocation, and knowledge distillation have been proposed to enhance the convergence, generalization, and overall performance of FL models in Non-IID environments [7].
(2) *Privacy Protection and Security Management*: FL is based on local computation and parameter exchange, ensuring that users' raw data always remains on their own devices and thus providing a preliminary structural data privacy protection. However, FL typically involves a large number of heterogeneous nodes that lack complete mutual trust, resulting in a highly decentralized and open system architecture. Under such circumstances, sensitive information may still be exposed through model inversion or inference attacks. Therefore, it is critical not only to safeguard the privacy and security of participants' data throughout the model training process, but also to achieve comprehensive security and trust management across the entire FL workflow. Specifically, a robust security framework for FL should be established on multiple privacy-preserving technologies, including local differential privacy, homomorphic encryption, and secure multi-party computation. Furthermore, by incorporating trust management mechanisms such as identity authentication and access control, the behaviors of participating nodes can be strictly regulated. This not only enhances the robustness of FL process but also ensures privacy and trustworthiness in collaborative model training.

1.1.3 Resource Scheduling Technologies

In end-edge-cloud networks, resource scheduling technologies serve as the fundamental guarantee for achieving multi-layer collaborative computing and efficient service delivery. With the rapid development of emerging technologies such as AI, end-edge-cloud resource scheduling is evolving from traditional static allocation toward adaptive and intelligent optimization. How to efficiently schedule multi-dimensional resources including computing, storage, and network to satisfy diverse service requirements has become particularly important.

Current research on resource scheduling primarily focuses on two major directions: On the one hand, from the perspective of algorithms (control), the emphasis is on developing traditional optimization algorithms (e.g., greedy and convex optimization algorithms), general heuristic and meta-heuristic algorithms (e.g., genetic algorithms (GAs) and particle swarm optimization), as well as learning-based intelligent methods to enhance decision-making capabilities. On the other hand, from the perspective of mechanisms (data), attention is given to the virtualization of underlying resources, network slicing, resource isolation, the design of time-sensitive protocols, and the perception-decision-action closed-loop mechanism in end-edge-cloud multi-layer collaboration. These efforts aim to ensure that resource scheduling strategies can be efficiently implemented and dynamically adjusted. In this part, we analyze the main research directions of resource scheduling technologies in end-edge-cloud networks from the perspectives of both algorithms and mechanisms.

Algorithms (Control): From the algorithmic perspective, the primary focus is on making resource scheduling decisions through mathematical modeling and intelligent algorithms to achieve efficient resource allocation and dynamic adjustment in complex environments. The main approaches can be categorized as follows:

(1) *Traditional Optimization Algorithms*: Traditional optimization algorithms, including linear programming, integer programming, and convex optimization, are grounded in rigorous mathematical modeling. They are capable of delivering provably optimal or high-quality solutions when the problem structure and constraints are explicitly defined and tractable. Li et al. [8] utilized continuous convex approximation to address the non-convex optimization problem in UAV-assisted edge computing networks, aiming to maximize energy efficiency. Du et al. [9] employed convex relaxation methods to solve a bilevel optimization problem, aiming to minimize the costs of both vehicular terminals and edge servers. Tran et al. [10] presented a collaborative resource allocation framework for multi-access edge computing (MEC) in 5G networks, utilizing mixed-integer linear programming (MILP) to jointly optimize user association, computing resource allocation, and task placement, thereby minimizing latency and energy consumption. However, the applicability of traditional optimization algorithms is often restricted by computational complexity and scalability issues in large-scale or highly dynamic end-edge-cloud environments.

(2) *Heuristic and Meta-heuristic Algorithms*: Heuristic and meta-heuristic algorithms have also been widely applied in task offloading and resource scheduling problems, due to their rapid-solving characteristics. Chai et al. [11] proposed a heuristic algorithm that successively solves computation offloading subproblem and resource allocation subproblem in cellular device-to-device (D2D) MEC systems by Kuhn–Munkres algorithm and Lagrangian dual method, respectively. Bi et al. [12] combined the genetic simulated annealing algorithm with particle swarm optimization to make computation offloading and resource scheduling

decisions, aiming to minimize energy costs. Udayakumar et al. [13] integrated optimal reinforcement learning with hybrid meta-heuristic algorithms to optimize resource allocation in IIoT systems. Aiming at addressing that allocating the same computing resources to different tasks may cause resource waste and make complex tasks difficult to converge to the optimal solution, [14] designed a multitask particle swarm optimization with a dynamic on-demand allocation strategy to dynamically allocate computing resources. Compared with traditional optimization algorithms, heuristic algorithms can efficiently search large solution spaces and quickly provide near-optimal solutions, even in the presence of heterogeneous resources and rapidly changing workloads.

(3) *Deep Reinforcement Learning (DRL) Algorithms*: In recent years, the application of AI in end-edge-cloud resource scheduling problems has increased. DRL algorithms have shown notable advantages in addressing resource scheduling challenges in dynamic environments [15–25]. DRL combines reinforcement learning principles with deep neural networks (DNNs) to determine optimal decision strategies [26]. By capturing the environment's dynamics and learning strategies through repeated interactions to achieve long-term goals, DRL effectively addresses challenges arising from the evolving heterogeneity of networks, mobile devices, and servers. Wang et al. [27] proposed an intelligent resource allocation algorithm based on dual attention DRL in a heterogeneous IIoT environment, which is used for selecting forwarding nodes and allocating channel access slots in specific network environments. Chen et al. [28] proposed a temporal attention deterministic policy gradient-based algorithm to address the joint optimization problem of computation offloading and resource allocation, where long short-term memory (LSTM) was integrated into the feature extraction network and a rank-based prioritized experience replay buffer was designed to accelerate and stabilize the convergence of model training. Tang et al. [29] combined LSTM, dueling deep Q-network, and double DQN techniques to address the problem of minimizing the expected long-term cost of MEC systems. In [30], Lu et al. defined data offloading in the low-earth orbit satellite-based Internet of Things (IoT) as a resource-constrained optimization problem. The optimization problem was formulated as a partially observable Markov decision process (MDP), and a novel multi-agent information broadcast and judgment algorithm was proposed to achieve rational resource scheduling.

Mechanisms (Data): From the perspective of mechanisms, the primary focus is on optimizing underlying network architectures and protocols to enhance the manageability and collaboration of resources across the end, edge, and cloud layers, thereby providing support for the effective implementation of upper resource scheduling strategies. Current research mainly revolves around the following directions:

(1) *Deterministic Scheduling Mechanisms*: Reliability and real-time has been key performance features of various communication networks. For example, in applications for autonomous driving, industrial robots, and high-definition live broadcast, highly reliable and timely data delivery is required to ensure excellent quality of service (QoS) [31]. Time-sensitive networks (TSN), developed from traditional industrial Ethernet, can provide deterministic communications for data flow transmission with predictable end-to-end latency and low packet loss rate in local area networks [32]. In essence, TSN constitutes a family of protocols that includes time synchronization, traffic scheduling, traffic shaping, and resource reservation and management. Time synchronization protocols (such as IEEE 802.1AS) adopt the precision time protocol (PTP)-based algorithms, providing a unified time reference for all scheduling and forwarding operations among network switching nodes. Traffic scheduling protocols (such as IEEE 802.1Qbv) introduce the time-aware shaper (TAS) to allow precise configuration of port gating times on Ethernet switches [33, 34]. Through gate control lists (GCL), these protocols regulate the priority queues that are allowed to be transmitted in each time slot, thus creating cyclic scheduling windows [35]. Traffic shaping protocols, as specified in standards such as IEEE 802.1Qav, IEEE 802.1Qcr, and IEEE 802.1Qch, are designed to regulate bandwidth usage, smooth traffic bursts, and reduce congestion throughout the network, thus supporting predictable and bounded end-to-end latency for time sensitive traffic [36]. Resource reservation and management protocols (such as IEEE 802.1Qat/802.1Qcc) employ centralized or distributed resource managers to perform advanced bandwidth allocation and path planning for critical traffic [37]. Through the stream reservation protocol (SRP), these standards dynamically register traffic information, ensuring that essential traffic receives the necessary end-to-end resources and is protected from contention with best-effort traffic.
Typical traffic shaping mechanisms, such as cyclic queuing and forwarding (CQF), provide each flow with deterministic per-hop latency by periodically alternating the receiving and sending states of ping-pong queues [38]. These ping-pong queues are deployed at each TSN switch port, operated by a pair of gates (i.e., a receiving gate and a sending gate) [39]. Gate operations are controlled by GCL to forward traffic, with the scheduling cycle serving as the operational period. Specifically, the scheduling cycle is divided into two time slots: even slots and odd slots. During even time slots, the receiving gate of queue Q_1 remains open to accept traffic from upstream nodes while its sending gate is closed; simultaneously, the sending gate of queue Q_2 is open to transmit traffic to downstream nodes while its receiving gate is closed. During odd time slots, the operations are reversed. In summary, the ping-pong queues enable deterministic transmission of traffic between any two adjacent switches through the cyclic alternation of working states.

(2) *Resource Virtualization and Network Slicing*: Given massive heterogeneous devices and diverse service requirements, how to flexibly manage and configure physical resources is a significant challenge. Resource virtualization technologies can decouple network functions from hardware via separating the data

plane and control plane, enabling logical abstraction and centralized control of resources, thereby endowing networks with flexible programmability. Taking network function virtualization (NFV) as the example, it focuses on software implementation of network services. Traditional networks rely on dedicated hardware devices to carry out critical functions such as routing, switching, and firewalls, resulting in rigid network deployment and difficulties in rapid innovation. NFV implements these network functions in software form running on general-purpose servers, leveraging virtual machines or container technologies for flexible deployment. This approach not only reduces equipment costs but also enables rapid adjustment of the scale and location of network functions according to traffic variations and service requirements [40, 41].

Based on resource virtualization, network slicing technologies further support logical partitioning and isolation of physical networks. The third generation partnership project (3GPP) defines each network slice as an end-to-end logical network comprising necessary network functions and resources, with each owning independent network topology, protocol stack, and security mechanisms, providing customized services for specific businesses [42–44]. Therefore, different kinds of data traffic can be transmitted smoothly on the same physical network without mutual interference [45–48]. Network slicing covers radio access networks (RAN), transport networks, and core networks, involving vertical integration from infrastructure to service applications [49]. Leveraging virtualization technologies, network slicing allows unified scheduling and management of multi-dimensional resources including computing, communication, and storage. Specifically, RAN slicing achieves flexible allocation of air interface resources through wireless resource scheduling and virtualization [50]; while transport network slicing implements traffic engineering in the data plane to guarantee bandwidth and latency requirements, and core network slicing enables on-demand deployment and elastic scaling of network elements through NFV [51]. Furthermore, network slicing supports end-to-end deployment across multiple operators, aiming to simultaneously accommodate diverse business scenarios including enhanced mobile broadband (eMBB), ultra-reliable low-latency communications (URLLC), and massive machine type communications (MMTC).

(3) *Communication and Computation Integration*: In traditional networks, communication and computation are generally separated, which represents network devices focus exclusively on data forwarding and transmission, while data processing and computing tasks are entirely performed by end devices or servers. However, with the rapid development of emerging applications such as AI and IoT, massive volumes of data require transmission across networks. Besides, scheduling massive concurrent computing traffic over a unified computing platform encounters numerous challenges. Hence, it is necessary to realize integrated management for communication and computing resources, thus facilitating efficient ubiquitous computation. The key lies in shifting computing capabilities from the edge devices into the network itself to achieve deep integration of communication and computation. Specifically, programmable data processing

units can be incorporated into network equipment such as switches and routers, enabling them to not only perform data forwarding but also undertake partial computing and storage tasks. This approach allows data to be processed in real time at network nodes proximate to the source, substantially reducing bandwidth occupation and latency associated with data transmission.

1.2 Challenges in End-Edge-Cloud Networks

As an important architecture for supporting the digital society and intelligent industry, end-edge-cloud networks is increasingly becoming an efficient collaborative paradigm that data processing and intelligent applications rely on. Compared with traditional centralized cloud computing, end-edge-cloud networks shows unique advantages in improving resource utilization and reducing service response latency. However, it is also accompanied by higher system complexity and diversified challenges. In practical applications, end-edge-cloud networks need to cope with efficient communication in multi-layer heterogeneous network and intelligent computing in dynamically changing scenarios at the same time. On the one hand, end-edge-cloud networks puts forward higher requirements on the latency, reliability, and security of network connections. The communication process involves multiple access technologies and complex data flows, which are susceptible to multiple effects such as latency, packet loss, congestion, and security threats. On the other hand, there are significant differences among terminal, edge, and cloud in terms of computing and storage capabilities, energy consumption constraints, and task allocation mechanisms. How to achieve efficient resource scheduling and collaborative computing has become the core problem of system design. In the following, we will elaborate on the major challenges confronted by end-edge-cloud networks from the communication and computation aspects.

1.2.1 Challenges in Communication

In end-edge-cloud networks, communication links cover multiple levels including terminal devices, edge nodes, and cloud data centers, covering a wide geographical area, diverse protocols, and complex environments. This multi-level and distributed network communication environment brings many challenges to the efficient and stable operation of the system, which are mainly reflected in the following aspects:

(1) *Co-transmission of Multiple Types of Traffic*: With the advent of the intelligence of everything (IoE) era, numerous end terminals are connected to the Internet, and various application services are continuously emerging, resulting in a surge in the types and scale of data traffic carried in networks. For example, in the IIoT scenario where large AI models are deployed, the end-edge-cloud

networks need to transmit not only model training parameters related to FL and DNN inference tasks, but also traditional industrial traffic (e.g. time-triggered traffic, event-driven traffic, and best-effort traffic) as well as burst safety-critical traffic. However, the available bandwidth resources in networks are limited. Particularly, during the process of large AI model training and inference, where the frequent uplink and downlink transmission of model parameters requires a large amount of wireless/wired bandwidth, it is very easy to cause communication bottlenecks when large-scale terminals participate in FL. In addition, the end-edge-cloud collaborative architecture makes the data transmission path complicated. Therefore, how to schedule appropriate network time slots and bandwidth for different types of traffic to avoid resource waste and network congestion, thereby achieving high-throughput and low-latency traffic transmission requires further investigation.

(2) *Controllable Latency and High Reliability Guarantee*: Emerging businesses (e.g., Metaverse, autonomous driving, intelligent manufacturing, and industrial control) have put forward diverse QoS requirements for transmission latency, jitter, packet loss, and bandwidth [52]. For example, during the FL process of industrial large AI models, reliable network transmission of model parameters is particularly crucial for accelerating FL convergence. For parameter distribution, if the global model parameters can be distributed to more end devices, the FL can make full use of geographically dispersed data to train the global model. For parameter aggregation, if more locally trained parameters can be uploaded to centralized servers, the global model can achieve faster convergence speed and higher accuracy, thereby promoting efficient industrial production. DNN inference tasks must be completed within strict latency constraints. Improper resource scheduling can lead to task accumulation and degraded service quality [53]. In addition, training sample collection may involve multiple video streams and image processing, which requires high network bandwidth. Traditional networks adopt best-effort transmission methods, which is difficult to satisfy the diverse requirements of different traffic while ensuring end-to-end reliable transmission. Deterministic networks (e.g., TSN) and deterministic transmission mechanisms have been studied to provide bounded latency and packet loss for data traffic. However, in scenarios with large-scale heterogeneous device access, resource reservation and queue scheduling become complex. How to design intelligent communication resource scheduling strategies to ensure different task priorities and service availability remains an open research challenge.

(3) *Device Heterogeneity and Network Status Dynamics*: In end-edge-cloud networks, the performance (e.g., computing capability and storage capacity) and access methods (e.g., wired and wireless) of communication devices are highly heterogeneous. Meanwhile, due to the frequent access of multiple terminals and edge nodes, as well as rapidly fluctuating traffic loads, both the network topology and link quality (including bandwidth, channel conditions, and interference levels) are subject to real-time variations. For example, during the FL process involving industrial large AI models, wireless channels in IIoT environments are highly susceptible to interference and noise, resulting in constantly changing

link availability. Hence, it is easy for participating nodes to upload slowly and drop offline, making it difficult to guarantee the reliability of inference tasks and model distribution, which in turn affects the efficiency of global FL. However, traditional static or centralized scheduling methods are difficult to address such a highly dynamic and distributed environment. Furthermore, the uneven distribution of communication resources across the entire network further complicates cross-domain collaborative scheduling. Therefore, how to achieve real-time perception of network status and improve the adaptability of scheduling algorithms to heterogeneous devices, has become a core problem that restricts the efficient operation of end-edge-cloud networks and the implementation of intelligent services.

1.2.2 Challenges in Computation

The computation challenges of end-edge-cloud networks are mainly due to the heterogeneity of computing resources, diverse requirements of computing tasks, and energy limitations. In this part, we will elaborate on the following four aspects:

(1) *Heterogeneous Computing Resources*: In end-edge-cloud networks, terminals, edge servers, and cloud platforms are highly heterogeneous in terms of computing capability, storage capacity, and energy consumption constraints [54]. Terminals are characterized by limited computing capability and high energy sensitivity, making them suitable for lightweight tasks. Edge servers, with moderate computing capability, are capable of handling aggregation, inference, and other tasks of intermediate complexity. In contrast, cloud platforms hold abundant resources, making them well-suited for large-scale training and global aggregation. To satisfy the diverse requirements for computing resources of different tasks (e.g., fast Fourier transform (FFT), local model training, federated aggregation, and DNN inference), it is essential to rationally partition and allocate tasks to appropriate computing devices based on task priority and current workload, preventing overload or prolonged idleness of individual devices, thereby enhancing the overall system throughput and stability [55]. In addition, the system needs to possess the ability to dynamically allocate multi-dimensional resources in real time, laying the foundation for the implementation of efficient computing collaboration and resource scheduling mechanisms.

(2) *High Dimensionality of Computing Resource Scheduling Decisions*: Due to the large number of edge devices and high concurrency of tasks, distributed computing paradigms present an enormous solution for task allocation and resource scheduling. With the requirements of computing tasks being diverse, the computing capabilities, network conditions, residual energy, and cache resources of different devices are in constant flux. For instance, the FL process for industrial large AI models puts high forward on timeliness and temporal consistency of training tasks. FL typically involves multiple rounds with long periods and

strong temporal dependencies. If training nodes have insufficient computing resources, extended training latency may slow down global convergence speed. Traditional task allocation methods struggle to adapt to highly dynamic environments with complex resource constraints, resulting in uneven computing resource scheduling, network congestion, and excessive latency. Furthermore, in high-dimensional decision spaces, decision variables such as model selection, cache management, and resource scheduling are coupled, and optimization objectives (e.g., latency, energy consumption, model accuracy) are multi-faceted, further exacerbating the difficulty of task allocation. Therefore, it is necessary to adopt intelligent algorithms such as DRL and evolutionary computing to support real-time decision-making in high-dimensional state space and improve the system's adaptability and global optimization capabilities. However, these approaches also bring additional challenges in terms of algorithm convergence speed, computing resource consumption, and deployment complexity.

(3) *Non-IID Data Distribution and Model Generalization Guarantee*: During the process of FL for industrial large AI models, models needs to learn from massive, multi-scenario, and diverse data to possess strong generalization capabilities. However, in practical deployment, the data collected from geographically distributed terminals often exhibits significant differences in sample quantity, class distribution, and feature space, showing a highly Non-IID characteristic. Consequently, if the data at individual node lacks representativeness, the model may fail to capture all possible scenarios, thereby undermining the global model's generalization performance. Meanwhile, FL usually relies on collaborative learning among different nodes, each with limited computing capability and independent management. When the computing resource allocation and participating node selection decisions are suboptimal, the key data on some nodes cannot be fully utilized, which further exacerbates the partiality of data distribution and imbalance in model training. Consequently, designing efficient aggregation and scheduling mechanisms that can dynamically perceive data distribution and computing resource status to maximize the model's adaptability to complex industrial applications is an urgent challenge in the field of distributed intelligent computing.

(4) *Energy Constraints*: In end-edge-cloud networks, terminals are typically energy-constrained devices. Take FL as an example, frequent data transmission (uplink and downlink) and model training may significantly deplete the devices' energy [56]. In practical scenarios, how to ensure the timeliness of communication while balancing energy consumption to prevent equipment from exiting training due to energy exhaustion is a realistic challenge that needs to be considered.

1.3 Issues to Be Studied in This Book

With the booming development of AI technologies, large AI models have shown broad application prospects in end-edge-cloud network scenarios, such as IIoT. For specific, IIoT is expected to achieve more accurate production control and decision support via constructing AI services, which is composed of the three phases: training, inference, and deployment. As shown in Fig. 1.2, to address the challenges faced during the above phases, the research issues of this book are developed from the following aspects:

Communication and Computation Integrated Network Architecture: With the advent of the Internet of everything (IoE) era, tremendous data is generated at the network edge, which needs transmitting and processing efficiently to provide various advanced services. For instance, in IIoT, with the collected multisensory data, the potential failures of robotic arms can be detected instantly via mathematical optimization methods. These services have common features, namely, computing-intensive and time-sensitive. To this end, numerous computing paradigms are proposed to achieve efficient computation in IoE scenarios. Specifically, cloud computing has powerful computing capabilities, yet long-distance data transmission suffers from a prohibitive response latency. To provide low-latency computing services, MEC arises as the computing resources sink from the cloud to the edge, such as BSs, industrial gateways (IGWs), or road side units. Moreover, taking into account data privacy laws and regulations, distributed computing has attracted considerable attention from both

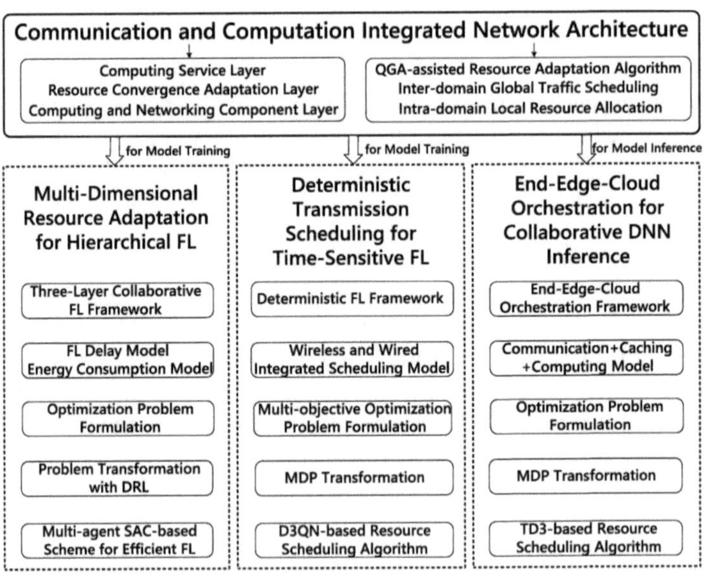

Fig. 1.2 An illustration of issues to be studied

academia and industry since the dispersed computing resources can be collaboratively utilized, e.g., FL. However, when the computing traffic tends to be diversified, customized, and intelligentized, single computing paradigm is difficult to satisfy the increasingly complex QoS and QoE requirements. Furthermore, computing task processing requires dynamic collaboration of multi-dimensional resources from both the communication and computation perspectives, which further complicates resource scheduling. In such cases, the relevant research content focuses on the following two aspects: (1) How to converge multiple computing paradigms into a unified computing platform and support integrated management for communication and computing resources, where massive concurrent computing traffic can be globally orchestrated in a more efficient manner and efficient ubiquitous computation can be achieved, is paramount important. (2) How to develop an intelligent resource adaptation algorithm and schedule massive concurrent computing traffic, to satisfy the differentiated and stringent requirements of computing traffic needs further investigation.

Multi-dimensional Resource Adaptation for Hierarchical FL: FL, as an emerging distributed learning paradigm, can avoid privacy leakage. However, optimizing FL in the distributed and resource-limited IIoT networks encounters several challenges. Firstly, due to excessive parameter transmission, traditional two-layer FL architecture may lead to high backbone communication overhead. Secondly, in the FL, the optimization objective is to minimize the FL evaluating loss, which is represented by an implicit function. The mapping relationship between the objective function and decision variables cannot be characterized accurately. As such, it is difficult to directly solve the problem via traditional optimization methods. Also, the decision variables of the device selection and resource allocation are correlated, which further complicates decision making process. In addition, the long-term energy consumption constraint couples to device selection and resource allocation decisions over time, and yet the decisions have to be made without foreseeing the future network dynamics. Inappropriate decision making may exhaust the on-device energy, consequently slowing down the FL convergence. To this end, the relevant issues are discussed from the following two aspects: (1) We consider a three-layer collaborative FL architecture empowered with a hierarchical aggregation algorithm to reduce transmitted model parameters between the edge and the cloud. (2) We investigate how to leverage deep RL approaches to dynamically select the participating devices and allocate cherished spectrum and computing resources for optimizing FL in the distributed IIoT networks.

Deterministic Transmission Scheduling for Time-Sensitive FL: FL is proposed to enable collaborative DNN model training among devices and a centralized server via periodically exchanging model parameters (i.e., FL traffic), for the purpose of achieving high learning accuracy with preserved data privacy and reduced communication overhead. Yet, operating FL over resource-constrained and time-sensitive IIoT faces a big challenge, that is, how to simultaneously transmit the FL traffic and traditional industrial traffic (e.g., time-triggered traffic, event-driven traffic, and best-effort traffic) over the same network and satisfy their diversified requirements.

In such a case, reliably aggregating the device-side model parameters in the centralized server is important to accelerate FL convergence. Therefore, the relevant research points are as follows: (1) We present a hierarchical deterministic FL framework for the time-sensitive IIoT, in which FL can efficiently operate with bounded end-to-end delay, jitter, and packet loss guarantee via temporal resource scheduling for the wireless and wired integrated TSN. (2) We propose a dueling double deep Q-network (D3QN) based resource scheduling algorithm to efficiently determine temporal resource allocation for FL traffic within the integrated TSN. After offline training, the proposed algorithm can facilitate FL convergence and adapt to network dynamics due to the traffic-by-traffic scheduling ability.

End-Edge-Cloud Orchestration for Collaborative DNN Inference: During the DNN inference phase, computing and caching resources are necessities for inference task execution. However, due to the heterogeneity of industrial facilities, dedicated pre-trained DNN models need to be provided for inference tasks based on the dynamic service requests, which poses a great challenge to efficiently manage multi-dimensional resources in IIoT networks to provide low-latency and high-accuracy inference services. To this end, the relevant research points are summarized as follows: (1) We introduce an end-edge-cloud orchestration architecture to flexibly coordinate the inference task assignment and DNN model placement, thereby providing low-latency DNN inference services in IIoT networks. (2) We formulate a joint task assignment and DNN placement (TADP) problem to maximize the average accuracy of massive inference tasks while satisfying the delay requirements under constrained spectrum, caching, and computing resources. (3) To effectively solve the joint optimization problem, we transform the TADP problem into an MDP, and propose a DRL-based algorithm to learn the inference task assignment and DNN model placement policy and make instant resource allocation decisions for DNN inference.

1.4 Overview of the Book

This book focuses on the intelligent resource scheduling in end-edge-cloud networks. An communication and computation integrated network architecture and multiple resource scheduling algorithms are proposed to facilitate efficient FL model training and real-time DNN inference, thereby supporting the full lifecycle of intelligent industrial applications. The remainder of this book is organized as follows.

In Chap. 2, we design a communication and computation integrated network architecture, named (Com)^{2}Net. Via (Com)^{2}Net, enormous computing traffic can be scheduled across space-air-ground domain, end-edge-cloud domain, and multi-data center domain, which facilitates ubiquitous connectivity and collaborative computation, thus supporting diverse advanced computing services. Then, open research issues that are fundamental for efficient collaborative computation in (Com)^{2}Net are discussed. Furthermore, aiming at satisfying numerous QoS and QoE requirements of diverse computing services, an intelligent resource adaptation scheme is proposed to dynamically orchestrate multi-dimensional resources for massive concurrent computing tasks, in which quantum genetic algorithms are utilized to make the best joint decision for inter-domain traffic scheduling and intra-domain resource allocation. Finally, we present a case study to evaluate the proposed QGA-based resource adaptation algorithm for optimizing hierarchical FL in the $(\text{Com})^2$Net.

In Chap. 3, to make the best joint decision of device selection and computing and spectrum resource allocation for optimizing FL performance in distributed IIoT networks, we introduce a three-layer collaborative FL architecture to support DNN training. Specifically, using the data dispersed in IIoT devices, the industrial gateways locally train the DNN model and the local models can be aggregated by their associated edge servers every FL epoch or by a cloud server every a few FL epochs for obtaining the global model. Then, we propose a "*reinforcement on federated*" (RoF) scheme, based on deep multi-agent reinforcement learning to solve the stochastic optimization problem with the objective of minimizing FL evaluating loss while satisfying delay and long-term energy consumption requirements. RoF scheme can facilitate efficient FL and achieve better performance compared with state-of-the-art benchmarks.

The FL is a long-term and iterative process, in which model parameters are interacted alternately among participating devices. Whether the model parameters can be reliably transmitted is especially important to accelerate FL convergence. Thus, in Chap. 4, we present a three-layer (i.e., device, field, and factory layers) deterministic federated FL framework, named DetFed, which accelerates collaborative learning process for ultra-reliable and low-latency IIoT via integrating 6G-oriented TSN. An integrated TSN is considered to establish connections among the three layers, where a CQF mechanism is deployed in each switch to support deterministic model parameter transmission with microsecond-level delay and near-zero packet loss requirements. Moreover, we propose a dynamic resource scheduling algorithm, based on DRL, to simultaneously maximize the scheduling success ratio of parameter flows and learning accuracy of FL models while satisfying the deterministic requirements of delay, jitter, and packet loss. Experimental results based on real-world dataset demonstrate that the proposed DetFed significantly accelerates FL convergence and improves learning accuracy as compared to state-of-the-art benchmarks.

To address the challenge to the resource-limited IIoT networks during the DNN inference phase, in Chap. 5, we introduce an end-edge-cloud orchestration architecture, in which the inference task assignment and DNN model placement are flexibly coordinated. Besides, aiming at achieving efficient DNN inference, a multi-dimensional resource management problem is formulated to maximize the average inference accuracy while satisfying the strict delay requirements of inference tasks. Due to the mix-integer decision variables, it is difficult to solve the formulated problem directly. Thus, we transform the formulated problem into an MDP and propose a DRL-based resource management scheme to make real-time optimal resource allocation decisions. Simulation results are provided to demonstrate that the proposed scheme can efficiently allocate the available spectrum, caching, and computing resources, and improve average inference accuracy by 31.4% compared with the DDPG benchmark.

Finally, we conclude this book and briefly discuss further research directions for resource scheduling in end-edge-cloud networks.

References

1. H. Hu, D. Wu, F. Zhou, X. Zhu, R.Q. Hu, H. Zhu, Intelligent resource allocation for edge-cloud collaborative networks: a hybrid DDPG-D3QN approach. IEEE Trans. Veh. Technol. **72**(8), 10696–10709 (2023)
2. W. Zhang, Y. He, T. Zhang, C. Ying, J. Kang, Intelligent resource adaptation for diversified service requirements in industrial IoT. IEEE Trans. Cogn. Commun. Netw. **11**(4), 2648–2661 (2025)
3. L. Zhu, S. Lin, F.R. Yu, Y. Li, Collaborative computing optimization in train-edge-cloud-based smart train systems using risk-sensitive reinforcement learning. IEEE Trans. Veh. Technol. **73**(3), 3129–3141 (2023)
4. W. Zhang, N. Tang, D. Yang, R. Guo, H. Zhang, X. Shen, Det(Com)^2: deterministic communication and computation integration toward AIGC services. IEEE Wireless Commun. **31**(3), 32–41 (2024)
5. S. Vahidian, M. Morafah, C. Chen, M. Shah, B. Lin, Rethinking data heterogeneity in federated learning: introducing a new notion and standard benchmarks. IEEE Trans. Artif. Intell. **5**(3), 1386–1397 (2024)
6. Y. Chen, Z. Wang, Y. Min, Z. Liu, Decentralized-voting-based federated learning framework for lightweight node selection in edge collaborative IoT. IEEE Internet Things J. **12**(12), 20272–20287 (2025)
7. X. Pang, J. Hu, P. Sun, J. Ren, Z. Wang, When federated learning meets knowledge distillation. IEEE Wireless Commun. **31**(5), 208–214 (2024)
8. M. Li, N. Cheng, J. Gao, Y. Wang, L. Zhao, X. Shen, Energy-efficient UAV-assisted mobile edge computing: resource allocation and trajectory optimization. IEEE Trans. Veh. Technol. **69**(3), 3424–3438 (2020)
9. J. Du, F.R. Yu, X. Chu, J. Feng, G. Lu, Computation offloading and resource allocation in vehicular networks based on dual-side cost minimization. IEEE Trans. Veh. Technol. **68**(2), 1079–1092 (2018)
10. T.X. Tran, A. Hajisami, P. Pandey, D. Pompili, Collaborative mobile edge computing in 5G networks: new paradigms, scenarios, and challenges. IEEE Commun. Mag. **55**(4), 54–61 (2017)
11. R. Chai, J. Lin, M. Chen, Q. Chen, Task execution cost minimization-based joint computation offloading and resource allocation for cellular D2D MEC systems. IEEE Syst. J. **13**(4), 4110–4121 (2019)
12. J. Bi, H. Yuan, S. Duanmu, M. Zhou, A. Abusorrah, Energy-optimized partial computation offloading in mobile-edge computing with genetic simulated-annealing-based particle swarm optimization. IEEE Internet Things J. **8**(5), 3774–3785 (2020)
13. K. Udayakumar, S. Ramamoorthy, Intelligent resource allocation in industrial IoT using reinforcement learning with hybrid meta-heuristic algorithm. Cybern. Syst. (2022)
14. H. Han, X. Bai, Y. Hou, J. Qiao, Multitask particle swarm optimization with dynamic on-demand allocation. IEEE Trans. Evol. Comput. **27**(4), 1015–1026 (2022)
15. Q. Ye, W. Shi, K. Qu, H. He, W. Zhuang, X. Shen, Joint RAN slicing and computation offloading for autonomous vehicular networks: a learning-assisted hierarchical approach. IEEE Open J. Veh. Technol. **2**, 272–288 (2021)
16. K. Kang, D. Ding, H. Xie, Q. Yin, J. Zeng, Adaptive DRL-based task scheduling for energy-efficient cloud computing. IEEE Trans. Netw. Serv. Manag. **19**(4), 4948–4961 (2021)

17. B. Hazarika, K. Singh, S. Biswas, C.-P. Li, DRL-based resource allocation for computation offloading in IoV networks. IEEE Trans. Ind. Inform. **18**(11), 8027–8038 (2022)
18. H. Li, K.D.R. Assis, S. Yan, D. Simeonidou, DRL-based long-term resource planning for task offloading policies in multiserver edge computing networks. IEEE Trans. Netw. Serv. Manag. **19**(4), 4151–4164 (2022)
19. C. Fang, Z. Hu, X. Meng, S. Tu, Z. Wang, D. Zeng, W. Ni, S. Guo, Z. Han, DRL-driven joint task offloading and resource allocation for energy-efficient content delivery in cloud-edge cooperation networks. IEEE Trans. Veh. Technol. **72**(12), 16195–16207 (2023)
20. S. Zhang, Z. Ni, L. Kuang, C. Jiang, X. Zhao, Load-aware distributed resource allocation for MF-TDMA ad hoc networks: a multi-agent DRL approach. IEEE Trans. Netw. Sci. Eng. **9**(6), 4426–4443 (2022)
21. G. Pang, W. Liu, Y. Li, B. Vucetic, DRL-based resource allocation in remote state estimation. IEEE Trans. Wireless Commun. **22**(7), 4434–4448 (2022)
22. H. Zhang, H. Wang, Y. Li, K. Long, A. Nallanathan, DRL-driven dynamic resource allocation for task-oriented semantic communication. IEEE Trans. Commun. **71**(7), 3992–4004 (2023)
23. Y. Chi, Y. Zhang, Y. Liu, H. Zhu, Z. Zheng, R. Liu, P. Zhang, Deep reinforcement learning based edge computing network aided resource allocation algorithm for smart grid. IEEE Access **11**, 6541–6550 (2022)
24. C. Fang, H. Xu, Y. Yang, Z. Hu, S. Tu, K. Ota, Z. Yang, M. Dong, Z. Han, F.R. Yu et al., Deep-reinforcement-learning-based resource allocation for content distribution in fog radio access networks. IEEE Internet Things J. **9**(18), 16874–16883 (2022)
25. H. Zhang, H. Liang, X. Hong, Y. Yao, B. Lin, D. Zhao, DRL-based resource allocation game with influence of review information for vehicular edge computing systems. IEEE Trans. Veh. Technol. (2024)
26. K. Arulkumaran, M.P. Deisenroth, M. Brundage, A.A. Bharath, Deep reinforcement learning: a brief survey. IEEE Signal Process. Mag. **34**(6), 26–38 (2017)
27. Y. Wang, F. Shang, J. Lei, Multi-granularity fusion resource allocation algorithm based on dual-attention deep reinforcement learning and lifelong learning architecture in heterogeneous IIoT. Inf. Fusion **99**, 101871 (2023)
28. J. Chen, H. Xing, Z. Xiao, L. Xu, T. Tao, A DRL agent for jointly optimizing computation offloading and resource allocation in MEC. IEEE Internet Things J. **8**(24), 17508–17524 (2021)
29. M. Tang, V.W. Wong, Deep reinforcement learning for task offloading in mobile edge computing systems. IEEE Trans. Mob. Comput. **21**(6), 1985–1997 (2020)
30. J. Lu, Q. Li, B. Guo, J. Li, Y. Shen, G. Li, H. Su, A multi-task oriented framework for mobile computation offloading. IEEE Trans. Cloud Comput. **10**(1), 187–201 (2019)
31. K. Zanbouri, M. Noor-A-Rahim, J. John, C.J. Sreenan, H.V. Poor, D. Pesch, A comprehensive survey of wireless time-sensitive networking (TSN): architecture, technologies, applications, and open issues. IEEE Commun. Surv. Tutor. **26**(1) (2024)
32. Y. Huang, S. Wang, T. Huang, Y. Liu, Cycle-based time-sensitive and deterministic networks: architecture, challenges, and open issues. IEEE Commun. Mag. **60**(6), 81–87 (2022)
33. T. Stüber, M. Eppler, L. Osswald, M. Menth, Performance comparison of offline scheduling algorithms for the time-aware shaper (TAS). IEEE Trans. Ind. Inform. **20**(7), 9736–9748 (2024)
34. Y. Ma, T. Zhang, X. Feng, P. Liu, H. Yang, F. Ren, Improving robustness of time-aware shaper in time-sensitive networking. IEEE Internet Things J. **12**(13), 25211–25223 (2025)
35. J. Ren, D. Yang, W. Zhang, K. Gong, W. Chen, W. Wu, H. Zhang, PTAS: PIFO-based time-aware shaper for massive concurrent flows in time-sensitive networks. IEEE Trans. Netw. Sci. Eng. **12**(1), 83–95 (2025)
36. D.A. Nascimento, S. Bondorf, D.R. Campelo, Modeling and analysis of time-aware shaper on half-duplex Ethernet PLCA multidrop. IEEE Trans. Commun. **71**(4), 2216–2229 (2023)
37. L. Zhao, P. Pop, S. Steinhorst, Quantitative performance comparison of various traffic shapers in time-sensitive networking. IEEE Trans. Netw. Serv. Manag. **19**(3), 2899–2928 (2022)
38. A. Nasrallah, A.S. Thyagaturu, Z. Alharbi, C. Wang, X. Shao, M. Reisslein, H. ElBakoury, Ultra-low latency (ULL) networks: the IEEE TSN and IETF DetNet standards and related 5G ULL research. IEEE Commun. Surv. Tutor. **21**(1), 88–145 (2018)

39. Y. Huang, S. Wang, B. Wu, T. Huang, Y. Liu, TACQ: enabling zero-jitter for cyclic-queuing and forwarding in time-sensitive networks, in *Proceedings of ICC 2021—IEEE International Conference on Communications*, pp. 1–6 (IEEE, 2021)
40. L. Zhao, J. Wang, J. Liu, N. Kato, Optimal edge resource allocation in IoT-based smart cities. IEEE Netw. **33**(2), 30–35 (2019)
41. R. Dong, C. She, W. Hardjawana, Y. Li, B. Vucetic, Deep learning for radio resource allocation with diverse quality-of-service requirements in 5G. IEEE Trans. Wireless Commun. **20**(4), 2309–2324 (2021)
42. I. Afolabi, T. Taleb, K. Samdanis, A. Ksentini, H. Flinck, Network slicing and softwarization: a survey on principles, enabling technologies, and solutions. IEEE Commun. Surv. Tutor. **20**(3), 2429–2453 (2018)
43. W. Wu, C. Zhou, M. Li, H. Wu, H. Zhou, N. Zhang, X.S. Shen, W. Zhuang, AI-native network slicing for 6G networks. IEEE Wireless Commun. **29**(1), 96–103 (2022)
44. M.K. Motalleb, V. Shah-Mansouri, S. Parsaeefard, O.L.A. López, Resource allocation in an open RAN system using network slicing. IEEE Trans. Netw. Serv. Manag. **20**(1), 471–485 (2022)
45. J. Wang, J. Liu, J. Li, N. Kato, Artificial intelligence-assisted network slicing: network assurance and service provisioning in 6G. IEEE Veh. Technol. Mag. **18**(1), 49–58 (2023)
46. H. Babbar, S. Rani, A.A. AlZubi, A. Singh, N. Nasser, A. Ali, Role of network slicing in software defined networking for 5G: use cases and future directions. IEEE Wireless Commun. **29**(1), 112–118 (2022)
47. T.K. Rodrigues, N. Kato, Network slicing with centralized and distributed reinforcement learning for combined satellite/ground networks in a 6G environment. IEEE Wireless Commun. **29**(1), 104–110 (2022)
48. Z. Wang, Y. Wei, F.R. Yu, Z. Han, Utility optimization for resource allocation in multi-access edge network slicing: a twin-actor deep deterministic policy gradient approach. IEEE Trans. Wireless Commun. **21**(8), 5842–5856 (2022)
49. W. Wu, N. Chen, C. Zhou, M. Li, X. Shen, W. Zhuang, X. Li, Dynamic RAN slicing for service-oriented vehicular networks via constrained learning. IEEE J. Sel. Areas Commun. **39**(7), 2076–2089 (2020)
50. S. Wijethilaka, M. Liyanage, Survey on network slicing for Internet of Things realization in 5G networks. IEEE Commun. Surv. Tutor. **23**(2), 957–994 (2021)
51. N. Van Huynh, D.T. Hoang, D.N. Nguyen, E. Dutkiewicz, Optimal and fast real-time resource slicing with deep dueling neural networks. IEEE J. Sel. Areas Commun. **37**(6), 1455–1470 (2019)
52. Y. Cai, J. Llorca, A.M. Tulino, A.F. Molisch, Ultra-reliable distributed cloud network control with end-to-end latency constraints. IEEE/ACM Trans. Netw. **30**(6), 2505–2520 (2022)
53. C. Yi, J. Cai, A truthful mechanism for scheduling delay-constrained wireless transmissions in IoT-based healthcare networks. IEEE Trans. Wireless Commun. **18**(2), 912–925 (2018)
54. Y. Wei, F.R. Yu, M. Song, Z. Han, Joint optimization of caching, computing, and radio resources for fog-enabled IoT using natural actor-critic deep reinforcement learning. IEEE Internet Things J. **6**(2), 2061–2073 (2018)
55. D. Qiao, M. Li, S. Guo, J. Zhao, B. Xiao, Resources-efficient adaptive federated learning for digital twin-enabled IIoT. IEEE Trans. Netw. Sci. Eng. (2024)
56. P. Dong, R. Shen, Q. Wang, Y. Zuo, Y. Li, D. Zhang, L. Zhang, W. Yang, Multipath TCP meets reinforcement learning: a novel energy-efficient scheduling approach in heterogeneous wireless networks. IEEE Wireless Commun. **30**(2), 138–146 (2022)

Chapter 2
Communication and Computation Integration Network Architecture

2.1 Introduction

With the advent of the IoE era, a wide variety of new network scenarios are emerging, such as Metaverse, ultrahigh-speed railway (USR), and intelligent IIoT [1]. Numerous end terminals are connected to the Internet, and tremendous data is generated at the network edge, which needs to be transmitted and processed efficiently [2]. As reported by Internet Data Center, the annual data generated worldwide will grow to 175 Zettabytes by 2025 [3]. Leveraging the largely increased data volume, various advanced services with different QoS and QoE requirements can be provided to improve operation and production efficiency [4]. For instance, in USR scenarios, with the captured audio and video data, the operation safety of USR vehicles can be guaranteed effectively via artificial intelligence algorithms [5]. In IIoT, with the collected multi-sensory data, the potential failures of robotic arms can be detected instantly via mathematical optimization methods [6]. These services have common features, namely, computation-intensive and time-sensitive [7]. To this end, how to facilitate efficient computation is especially critical to support abundant advanced services in IoE scenarios [8, 9].

Recently, numerous computing paradigms are proposed to achieve this goal [10]. Fueled by powerful computing capabilities, cloud computing significantly promotes the development of AI techniques [11–13]. Yet, long-distance data transmission suffers from a prohibitive response latency [14]. To provide low-latency computing services, multi-access edge computing arises as the computing resources sink from the cloud to the edge, such as base stations, industrial gateways, or road side units [15–19]. Moreover, taking into account data privacy laws and regulations, distributed computing has attracted considerable attention from both academia and industry since the dispersed computing resources can be collaboratively utilized, e.g., federated learning [20–25]. However, when the computing traffic tends to be diversified, customized, and intelligentize, single computing paradigm is difficult to satisfy the increasingly complex QoS and QoE requirements [26–30]. In such case, it is

© The Author(s), under exclusive license to Springer Nature Switzerland AG 2026
W. Zhang et al., *Intelligent Resource Scheduling in End-Edge-Cloud Networks*, Wireless Networks, https://doi.org/10.1007/978-3-032-07667-0_2

of paramount importance to converge multiple computing paradigms into a unified computing platform, where massive concurrent computing traffic can be globally orchestrated in a more efficient manner and the advanced service requirements can be guaranteed effectively.

Scheduling massive concurrent computing traffic over a unified computing platform encounters many challenges. *First*, computing task processing requires dynamic collaboration of multi-dimensional resources from both the communication and computation perspectives. Moreover, the unified computing platform needs to manage massive concurrent computing tasks generated from a wide variety of advanced services, such as large AI model training and image rendering, which further complicates computing traffic scheduling. Hence, it is necessary to design a novel architecture to support integrated management for communication and computing resources, thus facilitating efficient ubiquitous computation. *Second*, diverse computing services are fostering numerous new QoS and QoE requirements, such as deterministic latency and jitter, training accuracy, inference time and energy consumption, and data quality and privacy preserving [31]. Hence, it is essential to develop an intelligent resource adaptation algorithm to dynamically allocate multi-dimensional resources over the unified computing platform. To accelerate computation efficiency, how can we schedule massive concurrent computing traffic while satisfying the differentiated and stringent requirements needs further investigation.

In this chapter, we present a communication and computation integrated network architecture for massive concurrent computing traffic, named $(\text{Com})^2$Net. Enormous computing traffic can be scheduled across space-air-ground domain, end-edge-cloud domain, and multi-data center domain. Then, we propose a QGA-assisted scheme to support inter-domain traffic scheduling and intra-domain resource adaptation, thereby facilitating efficient multi-dimensional resource management. A detailed procedure of signaling interaction among end terminals, an inter-domain controller, intra-domain controllers, and computing platforms is presented. Finally, potential research directions such as computing-oriented routing protocol and blockchain-enabled computing transaction are introduced.

The remainder of this chapter is organized as follows. First, a communication and computation integrated network architecture is presented in Sect. 2.2, followed by some open research issues. Then, a quantum-assisted intelligent resource adaptation scheme is proposed in Sect. 2.3, followed by a case study in Sect. 2.3.4. Finally, we draw concluding remarks in Sect. 2.4.

2.2 $(\text{Com})^2$Net: Communication and Computation Integrated Network Architecture

As previously mentioned, how to satisfy diverse QoS and QoE requirements for advanced computing services remains a challenging issue [32]. Therefore, a multi-dimensional integrated system architecture for $(\text{Com})^2$Net is introduced to efficiently support traffic scheduling and resource adaptation for massive concurrent computing tasks. The architecture, as depicted in Fig. 2.1, aims to integrate ubiquitous connectivity and collaborative computation, while supporting a variety of computing services with distinct requirements.

Compared to conventional forwarding-dedicated networks, the (Com)^{2}Net is endowed with several brand-new features. For one thing, all network components are equipped with a computing unit to process computing tasks. Here, the network components refer to end devices, MEC servers, cloud servers, router devices, unmanned aerial vehicle (UAV), and satellites. Moreover, multiple computing centers, including data center, intelligent computing center, and supercomputer center, are embraced into the introduced architecture [34]. As such, enormous computing traffic can be scheduled across space-air-ground domain, end-edge-cloud domain, and multi-center domain, which facilitates ubiquitous connectivity and collaborative computation thus supporting diverse advanced computing services. For another thing, a two-level control mechanism is considered to jointly perform global and local management, respectively. In specific, to process the computing tasks, two optimal decisions should be dynamically made, i.e., *traffic scheduling* and *resource adaptation*. First, the traffic scheduling decision is to reasonably schedule computing tasks across the multiple

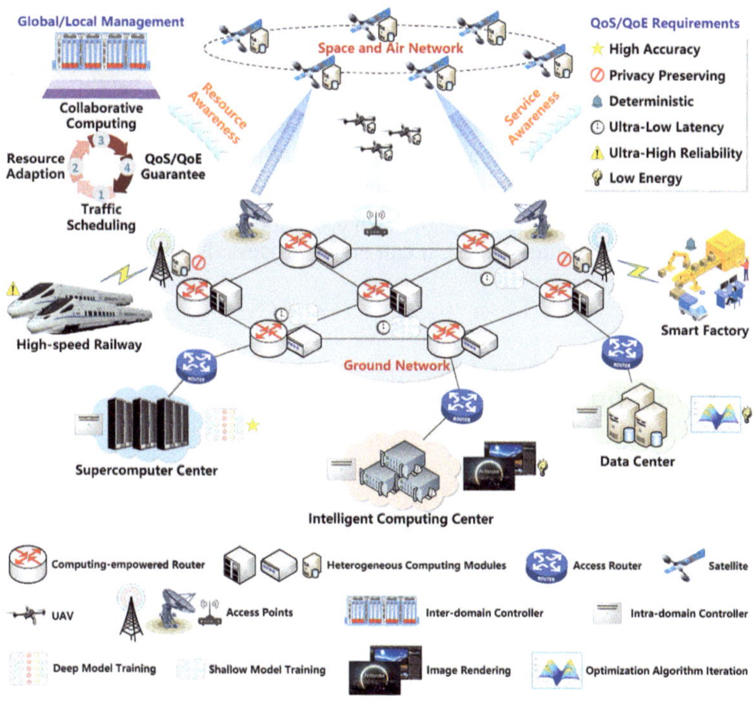

Fig. 2.1 An illustration of (Com)^{2}Net

domains. Second, the resource adaptation decision is to dynamically allocate multi-dimensional resources for transmitting and processing the corresponding computing tasks.

As shown in Fig. 2.2, the functional architecture of $(\mathrm{Com})^2$Net consists of three layers: computing and networking component layer, computing service layer, and resource convergence adaptation layer. In the following, we illustrate the detailed descriptions of each layer in the $(\mathrm{Com})^2$Net functional architecture.

2.2.1 Computing and Networking Component Layer

This layer plays a fundamental role for the entire $(\mathrm{Com})^2$Net architecture. Three types of components are considered to provide computing and networking services for computing tasks, i.e., network devices, computing devices, and integrated devices. Specifically, the network devices, such as routers, switches, and gateways, construct the core networks and are mainly responsible for hop-by-hop task forwarding. The computing devices consist of one or more processing modules such as central processing unit (CPU), graphics processing unit (GPU), and field programmable gate array (FPGA), and are in charge of instant task processing [35]. Compared to the network devices, the computing devices are generally distributed at the network edge, including multiple computing centers. That is to say, the computing tasks can be transmitted to computing devices by those network devices. In addition, the integrated devices converge the above two types of devices into one, thus providing forwarding and processing services for the computing tasks. For example, through introducing computing capabilities, the router devices can possess forwarding and computing functions, simultaneously. When a computing task is traversed through a computing-enhanced router device, it can either process the task locally or forward the task to other devices.

2.2.2 Computing Service Layer

In the computing service layer, the attribute features of various computing services can be abstracted according to their QoS and QoE requirements, and useful information can be obtained, including whether the task needs to be processed on multiple computing devices and how many computing devices need to be scheduled to complete the task. Such key information can provide service-level support for computing and networking convergence scheduling [36]. Furthermore, the computing tasks can be classified into different types, such as general computing tasks and intelligent computing tasks, which is helpful to conduct global task management. Typically, the computation process for general computing tasks cannot be split and needs to be carried out on a separate computing device. Taking fast fourier transform as an example, the calculation can be performed on an MEC server that has installed

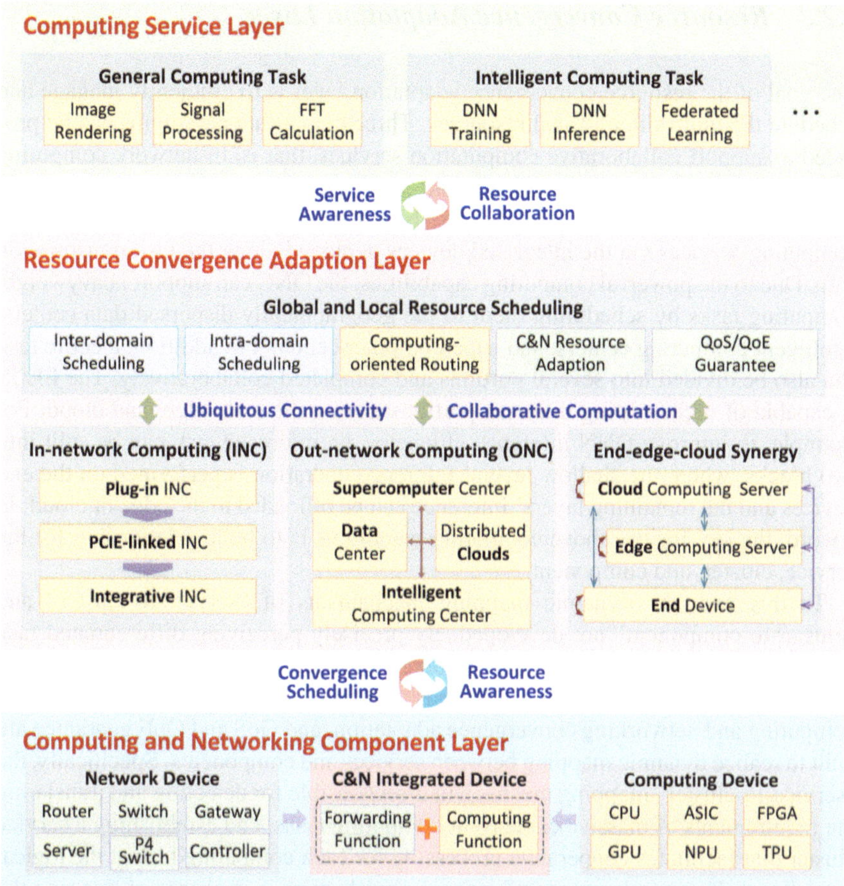

Fig. 2.2 (Com)^{2}Net functional architecture

the required mathematical tool. On the contrary, the computation process for intelligent computing tasks can be divided and coordinated among several computing devices. For instance, federated training requires DNN model frequently interacting among a centralized server and dispersed end devices. During the interaction process, multi-dimensional resources including networking, computing, and storage should be flexibly allocated to support efficient federated training.

2.2.3 Resource Convergence Adaptation Layer

The goal of the resource convergence adaptation layer is to efficiently manage and schedule the multi-dimensional resources. Three computing paradigms can be provided to support collaborative computation services, that is, in-network computing (INC), out-network computing (ONC), and end-edge-cloud synergy (EECS) computing [37]. For lightweight computing tasks, the INC paradigm can provide on-the-road computing services via the integrated devices deployed along the task transmission path. Due to the powerful computing capabilities, the ONC can support heavyweight computing tasks by scheduling them to the geographically dispersed data centers, intelligent computing centers, and supercomputer centers . In addition, a entire task can also be divided into several portions and completed collaboratively. The EECS is capable of orchestrating different portions among the end, edge, and cloud. For example, to improve DNN inference efficiency, an inference task can be split into two blocks, where the shallow layers' inference operation is performed on the end devices and the remaining layers' inference can be offloaded to the edge or cloud. In a word, the key for the above computing paradigms is to build relationship for the service, cluster, and component.

To this end, the dynamic mapping mechanisms of "service-to-cluster" and "cluster-to-component" are developed. By precisely perceiving differentiated task requirements and multi-dimensional resource attributes, functional modules including inter-domain scheduling, intra-domain scheduling, computing-oriented routing, computing and networking convergence adaptation, and QoS and QoE guarantee are built to realize dynamic mapping between services and components. Specifically, the "service-to-cluster" mapping mechanism is responsible for detecting and translating the requirements of massive concurrent computing tasks, and establishing a service cluster that facilitates cooperative processing for each computing task at the logical level. The "cluster-to-component" mapping mechanism is in charge of bringing the established service cluster into the physical computing and networking space and building the component cluster that performs the actual computing operations.

2.2.4 Open Research Issues for (Com)^{2}Net

In this part, we present several open research directions for (Com)^{2}Net.

The essential of traffic scheduling in (Com)^{2}Net is to forward a computing task to the most appropriate device and process it in an efficient way. This requires comprehensively take network states and available computing resource information into consideration, such that the optimal forwarding decision can be made to support traffic scheduling. Moreover, the availability of dispersed computing resources may dynamically change as the network topology reconstruction, which further complicates the forwarding decision making [40]. How to design computing-oriented routing protocols is an urgent issue to achieve ubiquitous connectivity and collaborative computation in (Com)^{2}Net.

New services, such as autonomous driving and metaverse, not only require ready-to-use computing resources, but also require deterministic guarantee of delay, jitter, and packet loss to satisfy their strict requirements [33]. For instance, image rendering computation of VR-based applications needs to be performed on the remote cloud, and users will suffer from vertigo and a significantly reduced QoE if the delay is too large. To effectively support advanced computing services, developing a deterministic scheduling mechanism for massive computing traffic is an important research issue.

With the development of blockchain technologies, dispersive and heterogeneous computing resources can be securely trade on a public platform [41]. Users can publish their requests for computing resources on the platform, and suppliers can sell or rent their available resources, thus realizing the schedulable, exchangeable, and on-demand resource supply. However, when misleading information is published, it will result in an imbalance between supply and demand, which is not conducive to efficient platform management. How to construct a trustworthy and transparent computing transaction platform is an interesting topic.

2.3 Intelligent Resource Adaptation for (Com)^{2}Net

In this section, we present a QGA-assisted resource adaptation algorithm and then investigate inter-domain and intra-domain scheduling schemes. In addition, we discuss the detailed procedure of signaling interaction in (Com)^{2}Net.

2.3.1 *QGA-Assisted Resource Adaptation Algorithm*

As illustrated in Fig. 2.1, two types of controllers are deployed in the (Com)^{2}Net architecture. One is the inter-domain controller, which is in charge of performing traffic scheduling across multiple domains from the global perspective. The other is the intra-domain controller, which is to allocate resources for computing tasks scheduled to the corresponding domains. To make the best decisions of traffic scheduling and resource allocation, each controller is endowed with a QGA-assisted resource adaptation algorithm [38]. To approach the optimal solution, the algorithm requires multiple iterations, including chromosome construction, initialization, selection, crossover, and mutation, and the iteration process is as follows [39].

(1) *Quantum Chromosome Construction and Population Initialization*: In the QGA-assisted algorithm, quantum chromosome i is utilized to represent individuals

in the population, i.e., $C_i = \{c_i^1, c_i^2, \ldots, c_i^D\}$. Here, D is the length of chromosomes, and $c_i^k \in [1, m]$, $1 \leq k \leq D$ denotes the traffic scheduling and resource allocation decisions (i.e., s_k and a_k) for kth computing task. Note that the alternative scheduling decisions have m cases. The chromosome is constructed using quantum bit (i.e., qubit) coding methods that differ from traditional genetic algorithms in the binary coding and integer coding methods. The qubit can exist in a superposition of states, thus enabling the chromosome simultaneously represent multiple solutions. Then, a set of population $P(0) = \{C_1^0, C_2^0, \ldots, C_p^0\}$ is initialized which corresponds to the potential solution for the joint traffic scheduling and resource allocation optimization problem, where p is the population size. In addition, a Hadamard's gate is utilized to carry out the transformation for each gene of the chromosomes, such that the chromosomes can be represented by a uniform superposition state.

(2) *Fitness Function Definition*: Each individual of the population is measured by calculating the corresponding value of fitness function Fit_i, which is used as the index for selecting the optimal individual. As such, individuals with low fitness will be excluded. If $M > f(s_k, a_k)$, $\text{Fit}_i = M - f(s_k, a_k)$; otherwise 0. Here, M is a constant value that ensures the fitness is non-negative, and $f(s_k, a_k)$ is the objective function of the algorithm. In addition, livability is utilized to evaluate the solution quality for genetic iteration process, i.e., $livability_i = \text{Fit}_i / \sum_{i=1}^{k} \text{Fit}_i$. Obviously, larger livability means a larger fitness value, namely a better individual. Furthermore, individual with a larger livability has more opportunities to regenerate in the next generation, otherwise is the opposite. Thus, it is important to formulate an appropriate fitness function that guarantees the feasibility of the candidate individuals.

(3) *Quantum Chromosome Selection, Crossover, and Mutation*: Taking the optimal individual of the current iteration as the evolutionary goal, the next generation population can be formed through a quantum gate operation that modifies the chromosome's qubit coding. Specifically, three processes are performed to approach the optimal solution: In quantum chromosome selection process, the tournament approach is applied to select parent groups for the next generation. First, $p/2$ individuals are randomly selected from the population, and then two individuals with the highest survivability are selected as parent group members. The process is repeated until the number of parental members reaches p. In quantum chromosome crossover process, quantum crossover operators are utilized to combine the quantum chromosomes of parent individuals, which enables the algorithm to explore a larger solution space by utilizing quantum principles (e.g., entanglement and superposition). In quantum chromosome mutation process, quantum gates (e.g., Pauli-X gate) are used to avoid local optima by inverting the amplitudes of quantum chromosomes. With the above processes, the next generation population of quantum chromosomes can be formed accordingly.

(4) *Optimal Solution Search and Iteration*: Repeating the above steps until reaching the iteration termination condition, and then the optimal solution for traffic scheduling and resource allocation can be obtained.

The difference between the traditional genetic algorithm and the QGA-assisted algorithm is mainly in population coding and evolutionary strategies. The fundamental idea behind the QGA's population coding method is to encode chromosomes using qubits and quantum superposition states, which allows each chromosome to simultaneously represent information from multiple states. In addition, the population is updated using a quantum rotating gate, and the evolution is directed by the knowledge of the optimal individual in the current iteration. During the iteration process, each qubit's superposition state gradually tends to a deterministic state and reaches convergence, and thus accomplishing the optimization goal. Due to the unique coding and updating techniques, the QGA-assisted algorithm outperforms classical genetic algorithms in terms of population diversity, convergence speed, and convergence accuracy.

2.3.2 *Inter-Domain Global Traffic Scheduling*

As shown in Fig. 2.3, deploying the QGA-assisted algorithm in the inter-domain controller, the traffic scheduling decision can be centrally made to support massive concurrent computing task scheduling among multiple computing domains, such as space-air-ground domain, cloud-edge-end domain, and multi-center domain. Notably, two or more domains can be simultaneously selected to provide computing services for the tasks that have extremely strict QoS or QoE requirements. Moreover, two additional strategies can be generated after the traffic scheduling decision is determined. On the one hand, the QGA-assisted algorithm will generate the computing-oriented routing strategies for each task with differentiated computing resource demands. Implementing the routing strategies to the corresponding network component clusters, the massive tasks can be forwarded to their destination computing node. On the other hand, reasonable resource allocation strategies for network bandwidth will be generated to facilitate efficient task transmission. With these decisions and strategies, the established system can achieve efficient collaborative computing.

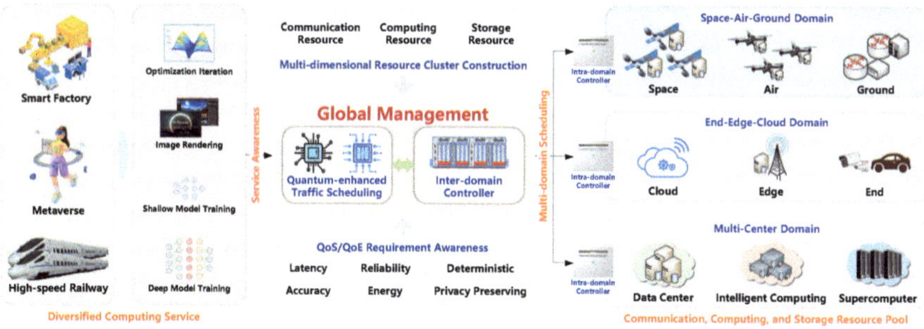

Fig. 2.3 Illustrate of inter-domain global scheduling in (Com)^{2}Net

2.3.3 *Intra-Domain Local Resource Allocation*

As shown in Fig. 2.4, with the QGA-assisted algorithm, the intra-domain controller can locally make resource allocation decisions for the computing tasks scheduled to the corresponding domain. Firstly, the resource information of the computing domain is dynamically telemetered, including the available computing and storage resources, remaining energy provision, and resource usage price. Secondly, heterogeneous computing and storage resources are collaboratively scheduled to support task processing. Two important decisions should be dynamically made, i.e., how many computing hardware units are selected to provide computing services? and how much computing resources for each hardware unit are allocated to process the current task? After sufficient iterations, the QGA-assisted algorithm is capable of extracting the dynamic task information and generating accurate resource allocation decisions in real time. As such, the QoS and QoE requirements of the computing task can be effectively guaranteed.

The QGA-assisted algorithm can perform inter-domain traffic scheduling and intra-domain resource allocation by offline optimization and online decision. The procedure of the QGA-assisted algorithm in $(\mathrm{Com})^2$Net involves the signaling interaction among computational users, an inter-domain controller, intra-domain controllers, and computing platforms. As illustrated in Fig. 2.5, the detailed procedure is as the following steps:

1. Computational users generate diverse computing tasks with differentiated requirements, and request moderate computing resources to complete the tasks.
2. The user information and service information are reported to the inter-domain controller to perform global management, such as traffic scheduling.
3. The inter-domain controller inquiries available resource information from the selected intra-domain controllers.

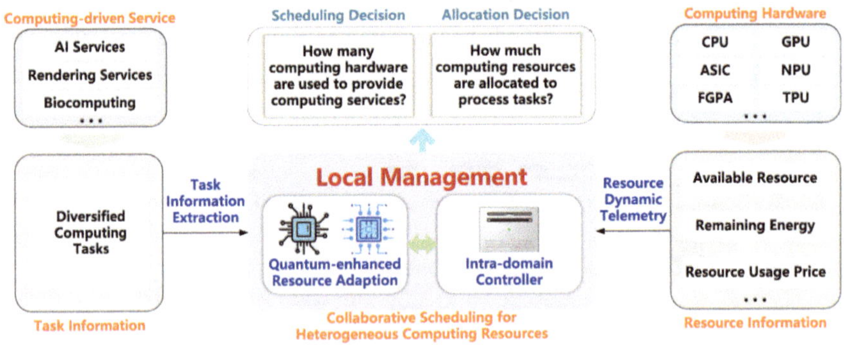

Fig. 2.4 Illustrate of intra-domain local scheduling in $(\mathrm{Com})^2$Net

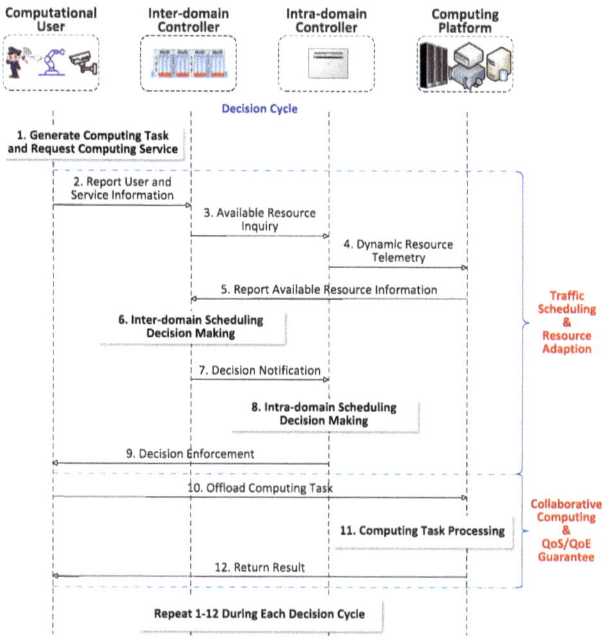

Fig. 2.5 Procedure of signaling interaction in (Com)^{2}Net

4. The intra-domain controllers collect the required available resource information from heterogeneous computing platforms via dynamic resource telemetry, including the available computing and storage resources, remaining energy provision, and resource usage price.
5. Reporting the required available resource information to the intra-domain controller, and further to the inter-domain controller.
6. The inter-domain controller runs the quantum-assisted traffic scheduling algorithms to make decisions with the obtained real-time information.
7. Deploying the determined traffic scheduling decisions (e.g., distributing computing tasks to corresponding domains) to the intra-domain controller.
8. The intra-domain controllers run the quantum-assisted resource adaptation algorithms to allocate computing and storage resources for the corresponding computing tasks.
9. Computational users enforce the received traffic scheduling and resource allocation decisions.
10. Computing service requests are supported using the allocated networking and computing resources. For instance, DNN inference tasks can be offloaded to MEC servers using spectrum resources.
11. Computing tasks are processed using the allocated computing resources. For example, federated learning tasks can be accelerated using GPU and FPGA resources.

12. Computing results are sent back to the initiated computational users for further utilization. For instance, an inference accuracy value of fault diagnostic applications can be sent back to an industrial controller, which can be utilized to determine whether to take maintenance measures for industrial equipment.

During the learning stage, steps 1–12 are repeated until the QGA-assisted algorithm reaches convergence. For each online decision cycle, steps 1–12 are disposable and have a quick response time [39]. Specifically, in offline optimization stage, the algorithm requires multiple iterations to approach the optimal solution, including chromosome construction, initialization, selection, crossover, and mutation, which is generally time consuming. In online decision stage, the obtained scheduling decisions can be deployed in the controllers to support the inter-domain traffic scheduling and intra-domain resource allocation which can be performed with a low latency [38].

2.3.4 Case Study

In this part, simulations are carried out to evaluate the proposed QGA-based resource adaptation algorithm for optimizing hierarchical FL in the $(\mathrm{Com})^2$Net.

We consider a multi-center integrated network for providing federated training services to intelligent industrial applications. Specifically, three data centers, two intelligent centers, and one supercomputer center are geographically distributed at different cities which construct a federated training-oriented $(\mathrm{Com})^2$Net. Each center involves a set of computing devices, e.g., CPU, GPU, and FPGA, which are connected into the $(\mathrm{Com})^2$Net via an access. When a smart factory connects the $(\mathrm{Com})^2$Net, enormous computing tasks have to be processed. We consider an accuracy-sensitive fault detection service, whose accuracy requirement is 90% for production safety. To facilitate efficient federated training, we adopt a three-layer learning architecture, in which a self-attention convolutional-based fault detection model is distributively trained on the centers' computing devices, and the updated model parameters are aggregated at the corresponding access routers every epoch or at a centralized factory server every a few epochs. In addition, we utilize a vibration signal-based dataset (i.e., *AWE dataset*[1]) to support the federated training for the model.

To accelerate FL convergence, we formulate the federated training as a joint bandwidth and computing resource optimization problem. Aiming at optimizing task completion ratio and testing accuracy, we utilize the proposed QGA-assisted algorithm to learn the resource adaptation policy. For performance comparison, we adopt a policy-gradient algorithm, named *double deep Q-network (DDQN)*, in which an online network and a target network are used to make resource allocation decisions. In this scheme, both online and target networks are fully-connected (FC) neural networks with two hidden layers, and the neuron numbers of the hidden layers are $[|S_{\mathrm{dim}}|, 1000, 500, |A_{\mathrm{dim}}|]$. In the simulation, the dimensions of network state $|S_{\mathrm{dim}}|$ and resource allocation decision $|A_{\mathrm{dim}}|$ are 688 and 4, respectively. In addition, the

[1] Online available. https://github.com/Intelligent-AWE/DeepHealth..

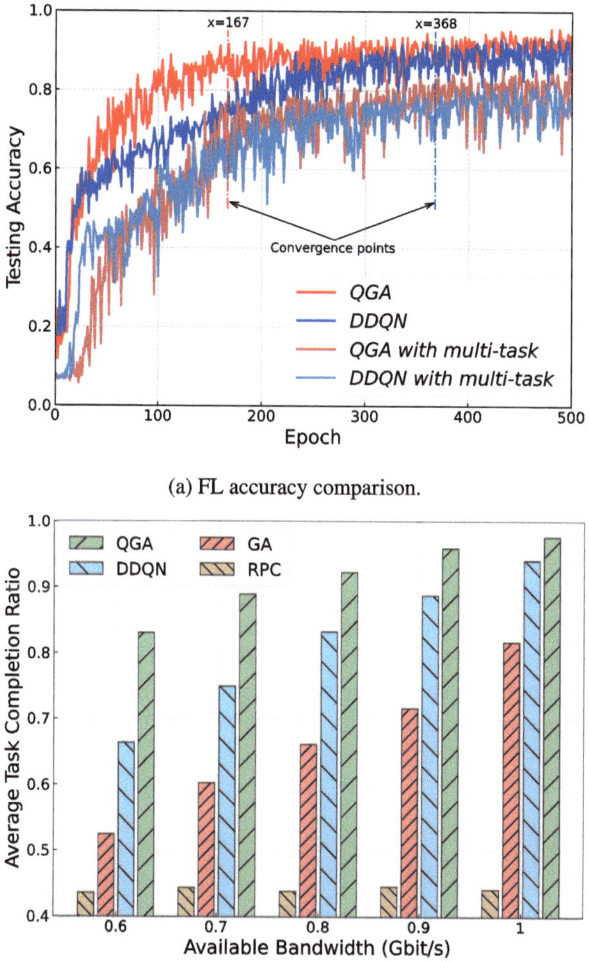

(b) Average task completion ratio comparison.

Fig. 2.6 Performance comparison between the proposed QGA-assisted scheme and DDQN benchmark

learning rate of this scheme is set to 1×10^{-4}. In addition, the genetic algorithm and random probabilistic configuration (RPC) are adopted as benchmark schemes.

We evaluate the performance of the proposed QGA-assisted resource adaptation scheme based on real-world AWE dateset. Figure 2.6a shows the convergence performance of the proposed scheme with respect to FL epochs. Three important observations can be obtained from the simulation results. First, when the available bandwidth is the same, the proposed scheme operated in a single-task scenario can obtain a higher accuracy than that in a multi-task scenario within 500 test epochs.

Second, the proposed scheme can reduce the number of training epochs until FL convergence by 54.62 and 12.5% in both scenarios compared with the DDQN scheme. The reason is that the quantum-based scheme can allocate network resources for FL traffic more reasonably via utilizing quantum superposition states, which allows each chromosome to simultaneously represent information from multiple states and thus obtaining the optimal resource allocation decision. Second, the proposed scheme achieves higher testing accuracy at the end of the FL training process in both single-task and multi-task scenarios. This indicates that the QGA-assisted scheme can support efficient transmission services for FL parameter aggregation and distribution within the $(\mathrm{Com})^2$Net.

As shown in Fig. 2.6b, we evaluate the average task completion ratio with respect to different available bandwidth resources. It can be seen that the proposed QGA-assisted scheme is capable of improving average task completion ratio compared with the benchmark schemes, which indicates that networking and computing resource adaptation are optimized. In addition, with the increase of the available bandwidth resources, the average task completion ratio achieved by the QGA-assisted scheme, DDQN, and GA increases. Particularly, when the amount of available bandwidth is 1 Gbit/s, the proposed scheme improves the average task completion ratio by 3.7, 19.5, and 121.6% compared with the DDQN, GA, and RPC benchmarks. The underlying reason is that more available bandwidth resources can be utilized to transmit FL model parameters over the $(\mathrm{Com})^2$Net, and the quantum superposition characteristic can enhance the decision-making capability of the proposed scheme for resource adaptation and thus improving the task completion ratio.

2.4 Summary

In this article, we have presented the communication and computation integrated network architecture, i.e., $(\mathrm{Com})^2$Net, for the efficient management of diverse advanced computing services and multi-dimensional resources. Furthermore, we have proposed a QGA-assisted resource adaptation scheme to facilitate inter-domain traffic scheduling and intra-domain resource allocation. By integrating space-air-ground domain, end-edge-cloud domain, and multi-data center domain, the $(\mathrm{Com})^2$Net can efficiently support collaborative computation for massive concurrent computing tasks. For the future work, we will study a quantum reinforcement learning based intent awareness scheme to enable a behavior-sensitive $(\mathrm{Com})^2$Net for metaverse-oriented computing service support.

References

1. D. Yang, W. Zhang, Q. Ye, C. Zhang, N. Zhang, C. Huang, H. Zhang, X. Shen, DetFed: dynamic resource scheduling for deterministic federated learning over time-sensitive networks. IEEE Trans. Mob. Comput. **23**(5), 5162–5178 (2024)
2. Q. Tang, F.R. Yu, R. Xie, A. Boukerche, T. Huang, Y. Liu, Internet of intelligence: a survey on the enabling technologies, applications, and challenges. IEEE Commun. Surv. Tutor. **24**(3), 1394–1434 (2022)
3. S. Duan, D. Wang, J. Ren, F. Lyu, Y. Zhang, H. Wu, X. Shen, Distributed artificial intelligence empowered by end-edge-cloud computing: A survey. IEEE Commun. Surv. Tutor. **25**(1), 591–624 (2022)
4. P. Zhang, Y. Zhang, N. Kumar, C.-H. Hsu, Deep reinforcement learning algorithm for latency-oriented IIoT resource orchestration. IEEE Internet Things J. **10**(8), 7153–7163 (2022)
5. Y. Wu, H.-N. Dai, H. Wang, Z. Xiong, S. Guo, A survey of intelligent network slicing management for industrial IoT: integrated approaches for smart transportation, smart energy, and smart factory. IEEE Commun. Surv. Tutor. **24**(2), 1175–1211 (2022)
6. N.H. Mahmood, G. Berardinelli, E.J. Khatib, R. Hashemi, C. De Lima, M. Latva-aho, A functional architecture for 6G special-purpose industrial IoT networks. IEEE Trans. Ind. Inform. **19**(3), 2530–2540 (2022)
7. B. Yin, J. Tang, M. Wen, Connectivity maximization in non-orthogonal network slicing enabled industrial Internet-of-Things with multiple services. IEEE Trans. Wirel. Commun. **22**(8), 5642–5656 (2023)
8. J. Huang, T. Yu, F. Yang, S. Zhang, W. Jiang, D. Niyato, AoI-aware resource allocation with interference avoidance for ultradense industrial Internet of Things networks. IEEE Internet Things J. **11**(17), 28787–28797 (2024)
9. D. Yang, A. Mahmood, S.A. Hassan, M. Gidlund, Guest editorial: industrial IoT and sensor networks in 5G-and-beyond wireless communication. IEEE Trans. Ind. Inform. **18**(6), 4118–4121 (2022)
10. J. Chen, J. Chen, K. Guo, Queue-aware service orchestration and adaptive parallel traffic scheduling optimization in SDNFV-enabled cloud computing. IEEE Trans. Cloud Comput. **11**(4), 3525–3540 (2023)
11. K. Peng, H. Huang, B. Zhao, A. Jolfaei, X. Xu, M. Bilal, Intelligent computation offloading and resource allocation in IIoT with end-edge-cloud computing using NSGA-III. IEEE Trans. Netw. Sci. Eng. **10**(5), 3032–3046 (2022)
12. H.R. Chi, M. de Fátima Domingues, H. Zhu, C. Li, K. Kojima, A. Radwan, Healthcare 5.0: in the perspective of consumer Internet-of-Things-based fog/cloud computing. IEEE Trans. Consum. Electron. **69**(4), 745–755 (2023)
13. T.H. Binh, H. Vo, B.M. Nguyen, H.T.T. Binh et al., Reinforcement learning for optimizing delay-sensitive task offloading in vehicular edge-cloud computing. IEEE Internet Things J. **11**(2), 2058–2069 (2023)
14. P. Lang, D. Tian, X. Duan, J. Zhou, Z. Sheng, V.C.M. Leung, Cooperative computation offloading in blockchain-based vehicular edge computing networks. IEEE Trans. Intell. Veh. **7**(3), 783–798 (2022)
15. Z. Sun, G. Sun, Y. Liu, J. Wang, D. Cao, BARGAIN-MATCH: a game theoretical approach for resource allocation and task offloading in vehicular edge computing networks. IEEE Trans. Mob. Comput. **23**(2), 1655–1673 (2023)
16. H. Zhou, Z. Wang, N. Cheng, D. Zeng, P. Fan, Stackelberg-game-based computation offloading method in cloud-edge computing networks. IEEE Internet Things J. **9**(17), 16510–16520 (2022)
17. W. Sun, Z. Li, Q. Wang, Y. Zhang, FedTAR: task and resource-aware federated learning for wireless computing power networks. IEEE Internet Things J. **10**(5), 4257–4270 (2022)
18. W. Zhou, J. Xia, F. Zhou, L. Fan, X. Lei, A. Nallanathan, G.K. Karagiannidis, Profit maximization for cache-enabled vehicular mobile edge computing networks. IEEE Trans. Veh. Technol. **72**(10), 13793–13798 (2023)

19. M. Babar, M.A. Jan, X. He, M.U. Tariq, S. Mastorakis, R. Alturki, An optimized IoT-enabled big data analytics architecture for edge-cloud computing. IEEE Internet Things J. **10**(5), 3995–4005 (2022)
20. W. Zhang, D. Yang, W. Wu, H. Peng, N. Zhang, H. Zhang, X. Shen, Optimizing federated learning in distributed industrial IoT: a multi-agent approach. IEEE J. Sel. Areas Commun. **39**(12), 3688–3703 (2021)
21. Q. Duan, J. Huang, S. Hu, R. Deng, Z. Lu, S. Yu, Combining federated learning and edge computing toward ubiquitous intelligence in 6G network: challenges, recent advances, and future directions. IEEE Commun. Surv. Tutor. **25**(4), 2892–2950 (2023)
22. Y. Liu, Y. Kang, T. Zou, Y. Pu, Y. He, X. Ye, Y. Ouyang, Y.-Q. Zhang, Q. Yang, Vertical federated learning: concepts, advances, and challenges. IEEE Trans. Knowl. Data Eng. **36**(7), 3615–3634 (2024)
23. L.T. Yang, R. Zhao, D. Liu, W. Lu, X. Deng, Tensor-empowered federated learning for cyber-physical-social computing and communication systems. IEEE Commun. Surv. Tutor. **25**(3), 1909–1940 (2023)
24. S. Zeng, Z. Li, H. Yu, Z. Zhang, L. Luo, B. Li, D. Niyato, HFedMS: heterogeneous federated learning with memorable data semantics in industrial metaverse. IEEE Trans. Cloud Comput. **11**(3), 3055–3069 (2023)
25. M. Hussain, S. Pal, Z. Jadidi, E. Foo, S. Kanhere, Federated zero trust architecture using artificial intelligence. IEEE Wirel. Commun. **31**(2), 30–35 (2024)
26. Y. Zhang, Q. Xu, L. Xu, C. Chen, X. Guan, Efficient flow scheduling for industrial time-sensitive networking: a divisibility theory-based method. IEEE Trans. Ind. Inform. **18**(12), 9312–9323 (2022)
27. Z. Zhao, G. Xu, N. Zhang, Q. Zhang, Performance analysis of the hybrid satellite-terrestrial relay network with opportunistic scheduling over generalized fading channels. IEEE Trans. Veh. Technol. **71**(3), 2914–2924 (2022)
28. C.-H. Hu, Z. Chen, E.G. Larsson, Scheduling and aggregation design for asynchronous federated learning over wireless networks. IEEE J. Sel. Areas Commun. **41**(4), 874–886 (2023)
29. Y. Zhu, B. Mao, N. Kato, A dynamic task scheduling strategy for multi-access edge computing in IRS-aided vehicular networks. IEEE Trans. Emerg. Top. Comput. **10**(4), 1761–1771 (2022)
30. Q. Tang, R. Xie, F.R. Yu, T. Chen, R. Zhang, T. Huang, Y. Liu, Distributed task scheduling in serverless edge computing networks for the Internet of Things: a learning approach. IEEE Internet Things J. **9**(20), 19634–19648 (2022)
31. Q. Ye, W. Shi, K. Qu, H. He, W. Zhuang, X. Shen, Joint RAN slicing and computation offloading for autonomous vehicular networks: a learning-assisted hierarchical approach. IEEE Open J. Veh. Technol. **2**, 272–288 (2021)
32. R. Xie, Q. Tang, Q. Wang, X. Liu, F.R. Yu, T. Huang, Satellite-terrestrial integrated edge computing networks: architecture, challenges, and open issues. IEEE Netw. Mag. **34**(3), 224–231 (2020)
33. D. Yang, K. Gong, J. Ren, W. Zhang, W. Wu, H. Zhang, TC-Flow: chain flow scheduling for advanced industrial applications in time-sensitive networks. IEEE Netw. Mag. **36**(2), 16–24 (2022)
34. L. Zeng, C. Yang, P. Huang, Z. Zhou, S. Yu, X. Chen, GNN at the edge: cost-efficient graph neural network processing over distributed edge servers. IEEE J. Sel. Areas Commun. **41**(3), 720–739 (2022)
35. S. Sundar, J.P. Champati, B. Liang, Multi-user task offloading to heterogeneous processors with communication delay and budget constraints. IEEE Trans. Cloud Comput. **10**(3), 1958–1974 (2020)
36. Q. Qi, X. Chen, A. Khalili, C. Zhong, Z. Zhang, D.W.K. Ng, Integrating sensing, computing, and communication in 6G wireless networks: design and optimization. IEEE Trans. Commun. **70**(9), 6212–6227 (2022)
37. T. Mai, H. Yao, S. Guo, Y. Liu, In-network computing powered mobile edge: toward high performance industrial IoT. IEEE Netw. Mag. **35**(1), 289–295 (2020)

38. M. Kim, S. Kasi, P.A. Lott, D. Venturelli, J. Kaewell, K. Jamieson, Heuristic quantum optimization for 6G wireless communications. IEEE Netw. Mag. **35**(4), 8–15 (2021)
39. T.Q. Duong, L.D. Nguyen, B. Narottama, J.A. Ansere, D. Van Huynh, H. Shin, Quantum-inspired real-time optimisation for 6G networks: opportunities, challenges, and the road ahead. IEEE Open J. Commun. Soc. **3**, 1347–1359 (2022)
40. J. Zhang, A. Sinha, J. Llorca, A.M. Tulino, E. Modiano, Optimal control of distributed computing networks with mixed-cast traffic flows. IEEE/ACM Trans. Netw. **29**(4), 1760–1773 (2021)
41. C. Zhang, M. Zhao, L. Zhu, W. Zhang, T. Wu, J. Ni, FRUIT: a blockchain-based efficient and privacy-preserving quality-aware incentive scheme. IEEE J. Sel. Areas Commun. **40**(12), 3343–3357 (2022)

Chapter 3
Multi-dimensional Resource Adaptation for Hierarchical Federated Learning

3.1 Introduction

With the wide deployment of IIoT, numerous devices are connected to the Internet, and voluminous industrial data is generated at the network edge [1–3]. As reported by Cisco, 2.4 TB data is burst out per minute from an industrial company in 2021 [3]. Utilized tremendous data, deep learning approaches, especially DNNs can facilitate industrial intelligence [4]. For example, a DNN model can be trained to provide fault diagnostic services for industrial facilities [5]. In industrial scenarios, different factories usually generate a huge amount of data that contains critical information, such as the operation conditions of industrial facilities. Since this data information generally plays a crucial role in safety pre-warning and fault diagnostic, it is of great significance to preserve the privacy of industrial data [6–8]. Traditional centralized training (CT) method needs to collect massive raw data from IIoT devices, e.g., industrial gateways, which leads to user privacy concerns [9]. Hence, taking into account devices' privacy-preserving requirements, the data would be better processed locally [10–12].

Federated learning, as an emerging distributed learning paradigm, can effectively solve the above issue [13]. Generally, FL operates over a two-layer architecture epoch-by-epoch. In each FL epoch, DNN models are locally trained at devices using their private dataset; and then the locally-trained model parameters are aggregated by a centralized server to update the global model parameters via a model aggregation algorithm, such as federated averaging (FedAvg) [14]. By running FL for multiple epochs until desired accuracy is achieved, a well-trained DNN model can be obtained to provide high-accuracy services. With the FL paradigm, DNN models can be efficiently trained based on the dispersed datasets held by devices without uploading raw data to a centralized server, thereby preserving privacy for the devices [15].

© The Author(s), under exclusive license to Springer Nature Switzerland AG 2026
W. Zhang et al., *Intelligent Resource Scheduling in End-Edge-Cloud Networks*, Wireless Networks, https://doi.org/10.1007/978-3-032-07667-0_3

Very recently, advanced FL schemes have been widely applied in wireless edge networks [16–20]. In particular, applying FL in IIoT can greatly improve industrial data utilization with preserved data privacy. To this end, we focus on facilitating efficient FL in distributed IIoT networks. Smart factories of an industrial company are generally located in different geographical areas. To provide services for all factories, such as intelligent fault diagnostic, the company requires a global DNN model to achieve satisfactory performance over the data generated by these factories. However, due to excessive parameter transmission, traditional two-layer FL architecture will lead to high backbone communication overhead.

Accordingly, we introduce a three-layer architecture, i.e., device-edge-cloud, to support FL in the considered distributed scenario, in which the industrial data held by IGWs are geographically dispersed among factories. In each epoch, a set of IGWs need to be selected to participate in the FL, i.e., *device selection*. DNN model parameters that are locally trained at the selected IGWs are aggregated by edge servers every FL epoch, and the edge aggregated model parameters are transmitted to the cloud server for further aggregation every a few FL epochs. In this way, the amount of transmitted model parameters between the edge and the cloud can be greatly reduced. To efficiently support local training and parameter transmission, on-device computing and network spectrum resources need to be judiciously allocated for the participating devices, i.e., *resource allocation*. Due to limited network resources, selecting all devices to participate in each epoch may result in a prohibitive delay for parameter uploading, which in turn consumes excessively high on-device energy. Hence, device selection and resource allocation should be jointly considered to optimize FL with satisfied delay and energy consumption requirements.

Optimizing FL in the distributed and resource-limited IIoT networks encounters several challenges. *First*, in the FL, the optimization objective is to minimize the FL evaluating loss, which is represented by an implicit function. The mapping relationship between the objective function and decision variables cannot be characterized accurately. As such, it is difficult to directly solve the problem via traditional optimization methods. Also, the decision variables of the device selection and resource allocation are correlated, which further complicates decision making process. *Second*, the long-term energy consumption constraint couples to device selection and resource allocation decisions over time, and yet the decisions have to be made without foreseeing the future network dynamics. Inappropriate decision making may exhaust the on-device energy, consequently slowing down the FL convergence. To address the above challenges, DRL is a feasible approach to make prophetic decisions through capturing network dynamics [21–23].

In this work, we investigate how to leverage deep RL approaches to dynamically select the participating devices and allocate cherished spectrum and computing resources for optimizing FL in the distributed IIoT networks. First of all, we formulate a joint device selection and resource allocation (JDSRA) problem with the objective of minimizing the FL evaluating loss while satisfying the epoch delay and energy consumption requirements. This problem cannot be directly solved via traditional optimization methods due to the implicit loss function of FL and the temporal correlation of energy consumption constraint. Therefore, we propose a multi-agent learning-based resource management scheme to solve the formulated problem. The learning agents can judiciously make the optimal device selection and resource allocation decisions to efficiently support the long-term FL process without violating the long-term energy consumption and strict delay constraints. The maximum entropy term is incorporated into the reward function of the learning algorithm to effectively improve the exploration. Through training the multi-agent algorithm centrally, the learning agents, deployed at edge servers, can cooperatively make the optimal decisions. Extensive simulation results demonstrate that the proposed scheme achieves better FL performance, as compared to the benchmarks. The main contributions of this chapter are summarized as follows:

- We introduce a three-layer collaborative FL architecture in the distributed IIoT networks, which can facilitate efficient FL while effectively reducing the backbone network traffic for DNN model parameter transmission;
- We formulate JDSRA as a stochastic optimization problem to optimize FL performance while satisfying the strict epoch delay and long-term energy consumption requirements, which is then transformed into a partially-observable Markov decision process (POMDP);
- We propose a "*reinforcement on federated*" (RoF) scheme, based on a deep multi-agent RL algorithm and a device refinement subroutine, to solve the POMDP. Through decentralized execution among multiple agents, the RoF scheme can cooperatively make the optimal decisions and accelerate convergence.

The remainder of this chapter is organized as follows. Section 3.2 reviews related works. In Sect. 3.3.1, a three-layer collaborative FL architecture is presented. Section 3.3 presents the system model, and then the problem is formulated and transformed in Sect. 3.4. Section 3.5 proposes the learning-based RoF scheme. Simulation results are given in Sect. 3.6, and the concluding remarks are provided in Sect. 3.7.

3.2 Related Work

3.2.1 FL in End-Edge-Cloud Networks

Recently, there are several research works on FL. A line of works focus on device selection to accelerate FL convergence. Generally, selecting more devices in each FL epoch can improve the convergence rate. To this end, Yang et al. [24] aimed at maximizing the number of participating devices in each FL epoch via exploiting superposition property of wireless multiple-access channels. Nevertheless, a large number of participating devices leads to prohibitive communication cost due to excessive parameter transmission. Wang et al. [25] proposed a learning-based scheme to flexibly select a set of appropriate devices, consequently speeding up FL convergence. Different from the above works, our work focuses on cooperatively selecting devices for three-layer FL architecture via multiple learning agents. In addition, a device refinement subroutine is adopted to obtain a refined set of selected devices, which can ensure the FL can be completed with satisfied delay and energy consumption requirements, thereby accelerating FL convergence.

3.2.2 *Resource Scheduling for FL*

Another line of works devote to resource allocation to facilitate efficient FL [26–32]. In [15], an FL delay minimization problem is studied in cellular networks, in which the spectrum resource is dynamically allocated for parameter transmission. In [33], an opportunistic spectrum access scheme is proposed to minimize the energy consumption of FL. The above works focus on implementing FL based on a traditional two-layer architecture. Different from these works, this chapter aims to optimize FL in a distributed IIoT network, in which the industrial data are dispersed geographically and backbone network is congested. In such case, instead of applying the two-layer architecture, we adopt a novel three-layer collaborative architecture to support FL, which can effectively reduce backbone network traffic for parameter transmission.

Deep RL techniques hold the potential to improve resource utilization in wireless networks via learning network dynamics [34–36]. Many advanced deep RL based algorithms have been developed to perform dynamic resource management [37–46]. In [47], a deep Q-network (DQN) based algorithm is proposed to decide the devices to conduct data collection in an unmanned aerial vehicle-assisted mobile edge network. In [21], Chen et al. adopted a double DQN algorithm to make computation offloading and packet scheduling decisions in radio access networks. Peng et al. [22] leveraged a DDPG-based algorithm to allocate spectrum resources in complex urban vehicular networks and learn the optimal task offloading policy. Moreover, aiming at optimizing the energy efficiency in a large-scale cognitive radio network, Kaur et al. [48] proposed a multi-agent Q-Learning based algorithm to obtain a decentralized radio resource allocation strategy. The above works provide valuable insights for achieving dynamic resource management in wireless networks. Different from the above works, we focus on facilitating efficient FL in distributed IIoT via leveraging deep RL approaches to judiciously allocate spectrum and computing resources.

In our preliminary work [49], we proposed a centralized resource management scheme, in which a single-agent based algorithm is designed to globally allocate network resources. As an extension of [49], in this chapter, we further propose a decentralized learning-based algorithm to perform judicious resource management for FL operation. In this algorithm, each geographically distributed base station is equipped with an edge controller that can only observe its individual environment information and make the decision for supporting FL independently.

3.3 System Model of Hierarchical FL

3.3.1 Three-Layer Collaborative FL Framework

To implement FL in distributed IIoT networks, we consider a three-layer collaborative FL architecture empowered with a hierarchical aggregation algorithm [50]. As shown in Fig. 3.1, the architecture is composed of the device, edge, and cloud layers endowed with different functionalities. The descriptions of each layer in the architecture are as follows.

- **Device Layer**: Energy-limited IGWs endowed with computing capability are geographically distributed in smart factories located in different cities. In each FL epoch, DNN models are locally trained at IGWs.
- **Edge Layer**: Multiple BSs are deployed to provide services for the IGWs. Each BS is equipped with an edge server to perform edge model parameter aggregation every FL epoch. Additionally, an edge controller deployed at each edge server can manage network resources and coordinate FL within the coverage of the edge server.
- **Cloud Layer**: A cloud server is deployed to perform global parameter aggregation, which is used to aggregate the updated parameters from edge servers after a few edge model parameter aggregations.

Within the three-layer architecture, DNN model parameters are infrequently transmitted between the edge and cloud, such that the backbone network traffic can be effectively reduced.[1]

In this architecture, we consider a set of edge servers, denoted by $\mathcal{M} = \{1, 2, \ldots, M\}$. Let $\mathcal{N}_m = \{1, 2, \ldots, |\mathcal{N}_m|\}$ be the set of IGWs within the coverage of edge server $m \in \mathcal{M}$. Let $\mathcal{D}_{m,n} = \{(\boldsymbol{x}_d, y_d)\}_d$ denote the local data set held by IGW n under edge server m, where $\boldsymbol{x}_d$ and y_d are the input and the label of each sample, respectively, and d is the sample index. Moreover, $\mathcal{D}^T_{m,n}$ and $\mathcal{D}^E_{m,n}$ are the training and evaluating data sets of the IGW, respectively, where $\mathcal{D}^T_{m,n} \cup \mathcal{D}^E_{m,n} = \mathcal{D}_{m,n}$. Note that the training data samples are assumed to follow a non-independent and identical distribution (non-IID). Similarly, $\mathcal{D}^T_m$ and $\mathcal{D}^E_m$ are the total training and evaluating data sets of the participating devices covered by edge server m, respectively, where $\mathcal{D}^T_m \cup \mathcal{D}^E_m = \mathcal{D}_m$. Notably, due to the data privacy-preserving requirement, the data samples held by each IGW cannot be shared with others during the FL process.

The operation procedure of the FL algorithm based on the three-layer architecture is given in Algorithm 1, which is detailed as follows.

- **Task Initialization (*Lines 1–2*)**: In the initialization stage, a training task is initiated at the cloud server, and a global DNN model with a set of initialized parameters

[1] The proposed three-layer architecture can also work in the case that not all the IGWs are able to connect the edge server. In such case, the locally trained model parameters can be directly transmitted to the cloud server for the global aggregation.

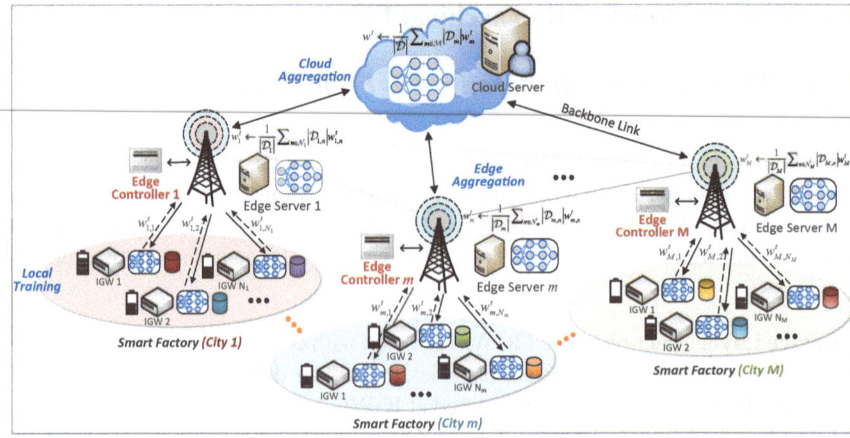

Fig. 3.1 The three-layer collaborative FL architecture for distributed IIoT networks

$\boldsymbol{w}_0$ is instantiated, such as a diagnostic DNN model that is utilized to classify the health conditions of industrial equipments.

- **Device Selection and Resource Allocation (*Lines 3–4*)**: Once the training task and the global DNN model are initiated, the edge controllers will make the device selection and resource allocation decisions for FL epoch $t \in \mathcal{R}$, respectively. In specific, the decisions include: (1) the set of IGWs that are selected to participate in the current FL epoch; and (2) the amounts of downlink/uplink spectrum and computing resources that are allocated to the selected IGWs. Given the decisions, a refined device set $\mathcal{N}_m^\star, \forall m \in \mathcal{M}$ can be obtained accordingly.
- **Global Model Parameter Distribution (*Lines 6–7*)**: With the refined device set $\mathcal{N}_m^\star, \forall m \in \mathcal{M}$, the global model parameters can be distributed from the edge or cloud to the participating IGWs.
- **Local Model Training (*Lines 8–12*)**: Once the global model parameters $\boldsymbol{w}_m$ are received, the participating IGWs will respectively perform model training on their local datasets for one FL epoch. Here, the parameters are updated via a stochastic gradient descend method, i.e.,

$$\boldsymbol{w}_{m,n} = \boldsymbol{w}_m - \lambda \nabla f(\boldsymbol{x}_d, y_d; \boldsymbol{w}_{m,n}), \forall d \in \mathcal{D}_{m,n}^T, \tag{3.1}$$

where λ is the learning rate, $\boldsymbol{w}_{m,n}$ is the updated parameters of each participating IGW, and $f(\boldsymbol{x}_d, y_d; \boldsymbol{w}_{m,n})$ is the loss on each training sample of the IGWs.
- **Edge Aggregation (*Lines 13–16*)**: When the local update is completed, each edge server synchronously aggregates model parameters from the participating IGWs at each FL epoch, i.e., $\mathrm{mod}(t, t_c) \neq 0$, and updates its global model parameter via the FedAvg algorithm, namely,

Algorithm 1: Three-layer FL algorithm

```
▷ Task initialization
Cloud: Initialize global DNN model parameter w_0;
▷ Device selection and resource allocation
Obtain a refined set of participating IGWs N*_m, ∀m ∈ M;
for each FL epoch t ∈ R do
    ▷ Model parameter distribution (downlink spectrum)
    Broadcast w_m to the selected IGWs;
    ▷ Local model training (on-device computing)
    for each BS m ∈ M in parallel do
        for each IGW n ∈ N*_m in parallel do
            for each training sample d ∈ D^T_{m,n} do
                w_{m,n} ← w_m − λ∇f(x_d, y_d; w_{m,n});
    ▷ Edge aggregation (uplink spectrum)
    if mod(t, t_c) ≠ 0 then
        w_m ← 1/|D^E_m| Σ_{m∈M,n∈N*_m} |D^E_{m,n}| w_{m,n};
        F(w) = F_e(w_m);
    ▷ Cloud aggregation (uplink spectrum)
    if mod(t, t_c) = 0 then
        w ← 1/|D^E| Σ_{m∈M} |D^E_m| w_m;
        F(w) = F_c(w);
return w, F(w).
```

$$\boldsymbol{w}_m = \frac{1}{\left|\mathcal{D}_m^E\right|} \sum_{m\in\mathcal{M},n\in\mathcal{N}_m^\star} \left|\mathcal{D}_{m,n}^E\right| \boldsymbol{w}_{m,n}, \tag{3.2}$$

where $\left|\mathcal{D}_{m,n}^E\right|$ and $\left|\mathcal{D}_m^E\right|$ are the numbers of evaluating samples held by each participating IGW and by the participating IGWs of edge server m. In the next FL epoch, the aggregated model parameters on each edge server can be distributed to the newly participating IGWs for further update. In such case, the evaluating loss of the DNN model can be described as

$$F_e(\boldsymbol{w}_m) = \frac{1}{M} \sum_{m\in\mathcal{M}} F_m(\boldsymbol{w}_m), \tag{3.3}$$

where

$$F_m(\boldsymbol{w}_m) = \frac{1}{\left|\mathcal{D}_m^E\right|} \sum_{d\in\mathcal{D}_m^E} f(\boldsymbol{x}_d, y_d; \boldsymbol{w}_m) + \frac{\mu}{2}\|\boldsymbol{w}_m\|^2, \tag{3.4}$$

where $f(\boldsymbol{x}_d, y_d; \boldsymbol{w}_m)$ denotes the loss on each evaluating sample of the participating IGWs with parameters $\boldsymbol{w}_m$ aggregated by edge server m, and $\mu\|\boldsymbol{w}_m\|^2/2$ is the regularization term that is added to prevent parameter overfitting, and μ is the regularization coefficient.

- **Cloud Aggregation (*Lines 17–20*)**: The cloud server aggregates model parameters from edge servers every t_c FL epochs, i.e., when $\text{mod}(t, t_c) = 0$, and updates the global model via the FedAvg algorithm, i.e.,

$$\boldsymbol{w} = \frac{1}{\left|\mathcal{D}^E\right|} \sum_{m \in \mathcal{M}} \left|\mathcal{D}_m^E\right| \boldsymbol{w}_m. \tag{3.5}$$

The evaluating loss is given by

$$F_c(\boldsymbol{w}) = \frac{1}{|\mathcal{D}|} \sum_{d \in \mathcal{D}} f(\boldsymbol{x}_d, y_d; \boldsymbol{w}) + \frac{\mu}{2} \|\boldsymbol{w}\|^2, \tag{3.6}$$

where $f(\boldsymbol{x}_d, y_d; \boldsymbol{w})$ represents the evaluating loss on each evaluating sample of the participating IGWs with parameters $\boldsymbol{w}$ aggregated by the cloud server, and $|\mathcal{D}|$ denotes the total number of evaluating samples held by all participating IGWs in the current FL epoch.

Note that when the connection between an IGW and the corresponding edge server is established, the IGW can send its information, including the number of training samples held by the IGW, to the edge server, and then such information is further reported to the centralized node (i.e., the cloud server). As such, the cloud server can have the information of the numbers of training samples held by distributed IGWs.

3.3.2 FL Delay Model

In this subsection, we present the FL delay and energy consumption models. Then, the FL evaluating loss is defined to measure the FL performance.

As previously mentioned, the FL process operates epoch-by-epoch. To facilitate efficient FL, each epoch is constrained by a delay requirement $T^{\max}$, and the total delay of each IGW during the FL operation is constrained by a budget $T_{m,n}^{\max}$. In specific, the delay for FL includes three parts, which are defined as follows. In FL epoch t, a certain number of IGWs needs to be selected to participate in the FL. Let $o_{m,n,t} \in \{0, 1\}, \forall m \in \mathcal{M}, n \in \mathcal{N}_m$ be the binary decision variable of device selection. Specifically, if $o_{m,n,t}$ equals 1, IGW n under edge server m is then selected, otherwise 0. We adopt orthogonal frequency allocation to support model parameter distribution and aggregation, which are operated with the available bandwidth $B^{\downarrow}$ (downlink) and $B^{\uparrow}$ (uplink), respectively. Here, the underlying assumption is that we do not consider other traffic loads in the considered scenario. This assumption is widely adopted in performance analysis of FL frameworks [12].

Thus, the downlink spectrum efficiency for edge server m is given by $\gamma_{m,n,t}^{\downarrow} = \log_2\left(1 + p_{m,t}|g_{m,n,t}|^2/\sigma^2\right), \forall m \in \mathcal{M}, n \in \mathcal{N}_m$, where $p_{m,t}$ is the transmission power, $\left|g_{m,n,t}\right|^2$ is the channel gain between edge server m and IGW n, and σ^2 is the

Gaussian noise power [51]. A fraction $\xi^{\downarrow}_{m,n,t}$ of bandwidth $B^{\downarrow}$ is allocated to IGW n from edge server m. Here, we have $0 \leq \xi^{\downarrow}_{m,n,t} \leq 1$, and $\sum_{m\in\mathcal{M},n\in\mathcal{N}_m} o_{m,n,t}\xi^{\downarrow}_{m,n,t} = 1$. Note that this chapter focuses on the spectrum allocation among IIoT devices, i.e., what percentage of resources should be allocated to an IIoT device. As such, we adopt a fractional bandwidth model. Hence, the downlink transmission rate of edge server m for IGW n is $R^{\downarrow}_{m,n,t} = \xi^{\downarrow}_{m,n,t} B^{\downarrow} \gamma^{\downarrow}_{m,n,t}$, $\forall m \in \mathcal{M}, n \in \mathcal{N}_m$.

When distributing model parameters to IGW n, the transmission delay $d^{\downarrow}_{m,n,t}$ can be calculated based on the following two cases: (1) $d^{\downarrow}_{m,n,t} = S_g/R^{\downarrow}_{m,n,t} + S_g/R^c$, when the global model parameters are distributed from the cloud server, where S_g (in bits) is the data size of the global DNN model parameter and R^c represents the backhaul transmission rate between the cloud and the edge; and (2) $d^{\downarrow}_{m,n,t} = S_g/R^{\downarrow}_{m,n,t}$, when the aggregated model parameters are distributed from edge servers to its covered IGWs.

With the orthogonal frequency allocation mechanism, the delay for distributing model parameters is determined by the slowest one, namely,

$$T^{\downarrow}_{t,\text{com}} = \max_{m\in\mathcal{M},n\in\mathcal{N}_m} d^{\downarrow}_{m,n,t}, \tag{3.7}$$

where $\max\{\cdot\}$ is the function to return the largest value.

In the considered IIoT networks, the distributed IGWs are endowed with a certain on-device computing capabilities, denoted by f (CPU cycles per second). The IGW's computing resource should be judiciously allocated to ensure that local model training can be completed within the strict delay threshold. Let $\eta_{m,n,t}$ be the decision variable of the fraction of computing resource the IGW used in FL epoch t, where $0 \leq \eta_{m,n,t} \leq 1$. Thus, the delay for local model training can be given by

$$\tau_{m,n,t} = \frac{a\left|\mathcal{D}^T_{m,n}\right|}{\eta_{m,n,t} f}, \forall m \in \mathcal{M}, n \in \mathcal{N}_m, \tag{3.8}$$

where a is the number of CPU cycles required for processing per bit data, and $\left|\mathcal{D}^T_{m,n}\right|$ is the size of training set held by each participating device. For different FL models, parameter a is set accordingly to reflect the effect of the adopted models on the training delay and energy consumption. Similarly, the delay for local model training is determined by the slowest one, i.e.,

$$T_{t,\text{cmp}} = \max_{m\in\mathcal{M},n\in\mathcal{N}_m} \tau_{m,n,t}. \tag{3.9}$$

Similar to the model parameter distribution, the uplink spectrum efficiency for parameter aggregation can be expressed as $\gamma^{\uparrow}_{m,n,t} = \log_2\left(1 + p_{m,n,t}|g_{m,n,t}|^2/\sigma^2\right)$, $\forall m \in \mathcal{M}, n \in \mathcal{N}_m$, where $p_{m,n,t}$ is the transmission power from IGW n to edge server m [52]. Let $\xi^{\uparrow}_{m,n,t}(0 \leq \xi^{\uparrow}_{m,n,t} \leq 1)$ denotes a decision variable of the fraction of uplink bandwidth $B^{\uparrow}$ allocated to IGW n, where $\sum_{m\in\mathcal{M},n\in\mathcal{N}_m} o_{m,n,t}\xi^{\uparrow}_{m,n,t} = 1$. Hence, the uplink transmission rate can be described as $R^{\uparrow}_{m,n,t} = \xi^{\uparrow}_{m,n,t} B^{\uparrow} \gamma^{\uparrow}_{m,n,t}, \forall m \in \mathcal{M}, n \in \mathcal{N}_m$. The transmission delay $d^{\uparrow}_{m,n,t}$ of uploading DNN model parameters from IGW n to edge server m also has two cases: (1) for the edge aggregation, the updated model parameters of IGWs need to be aggregated by the correlated edge server, and $d^{\uparrow}_{m,n,t} = S_l / R^{\uparrow}_{m,n,t}$; and (2) for the cloud aggregation, $d^{\uparrow}_{m,n,t} = S_l / R^{\uparrow}_{m,n,t} + S_l / R^c$. Here, S_l (in bits) is the parameter size of the local DNN model. Since the model parameter size is the same during the FL operation, we have $S_l = S_g$. In addition, DNN model parameters are infrequently transmitted between the edge and the cloud, thus we assume the backhaul transmission rates R^c in uplink and downlink are the same. Due to the synchronous aggregation mechanism, the delay for distributing model parameters is determined by the slowest IGW, i.e.,

$$T^{\uparrow}_{t,\text{com}} = \max_{m \in \mathcal{M}, n \in \mathcal{N}_m} d^{\uparrow}_{m,n,t}. \tag{3.10}$$

Taking $T^{\downarrow}_{t,\text{com}}$, $T_{t,\text{cmp}}$, and $T^{\uparrow}_{t,\text{com}}$ into consideration, the total delay for each FL epoch can be given by

$$T_t = T^{\downarrow}_{t,\text{com}} + T_{t,\text{cmp}} + T^{\uparrow}_{t,\text{com}}. \tag{3.11}$$

In addition, given the model parameter distribution, local training, and parameter uploading delays, the cumulative FL delay of each IGW is described as

$$T_{m,n,t} = T_{m,n,t-1} + d^{\downarrow}_{m,n,t} + \tau_{m,n,t} + d^{\uparrow}_{m,n,t}, \tag{3.12}$$

which is used to calculate the remaining time of FL, and thus affecting whether the IGWs can participate in the subsequent FL epochs. Here, $T_{m,n,t-1}$ denotes the cumulative FL delay of each IGW until epoch $t-1$.

3.3.3 Energy Consumption Model

In the long-term FL process, IGWs are constrained by their own energy capacity $E^{\max}_{m,n}$, which should be satisfied if the IGW is selected to participate in the FL. In specific, the energy consumption for FL includes two parts, which are calculated as follows, respectively. In the model parameter uploading stage, the energy consumption of IGW n covered by edge server m is given by

$$E^{\text{com}}_{m,n,t} = p_{m,n,t} d^{\uparrow}_{m,n,t}, \forall m \in \mathcal{M}, n \in \mathcal{N}_m. \tag{3.13}$$

Performing local model training at IGWs also consumes energy, which is affected by the allocated computing resource of the participating IGWs and the computing workload of the local training process. Hence, the energy consumption of local training at each IGW can be given by

$$E^{\mathrm{cmp}}_{m,n,t} = \frac{\kappa}{2} a \left|\mathcal{D}^{T}_{m,n}\right| (\eta_{m,n,t} f)^2, \forall m \in \mathcal{M}, n \in \mathcal{N}_m, \tag{3.14}$$

where κ is the effective capacitance coefficient of the computing chipset of IGWs [50].

Taking the above two parts into consideration, the cumulative energy consumption can be calculated by

$$E_{m,n,t} = E_{m,n,t-1} + E^{\mathrm{com}}_{m,n,t} + E^{\mathrm{cmp}}_{m,n,t}, \forall m \in \mathcal{M}, n \in \mathcal{N}_m, \tag{3.15}$$

which is utilized to calculate the remaining energy resource of IGWs, and hence evaluating whether the IGWs satisfy the long-term energy consumption requirements. Here, $E_{m,n,t-1}$ denotes the cumulative energy consumption of each IGW until epoch $t-1$.

In the considered architecture, FL operates epoch-by-epoch with the selected IGWs, which can satisfy the FL epoch delay requirement $T^{\max}$, time budget $T^{\max}_{m,n}$, and energy consumption constraint $E^{\max}_{m,n}$. The iteration process aims to obtain the optimal model parameters $\boldsymbol{w}^*$ that can minimize the evaluating loss on the evaluating datasets of the participating IGWs. To optimize FL under both the edge and cloud aggregation cases, the FL evaluating loss can be defined as

$$F(\boldsymbol{w}) = \begin{cases} F_e(\boldsymbol{w}_m), & \text{if } \mathrm{mod}(t, t_c) \neq 0; \\ F_c(\boldsymbol{w}), & \text{otherwise.} \end{cases} \tag{3.16}$$

If $\mathrm{mod}(t, t_c) \neq 0$, edge servers aggregate the covered model parameters from the participating IGWs, respectively, and obtain the evaluating loss $F_e(\boldsymbol{w}_m)$ with the aggregated parameters $\boldsymbol{w}_m, \forall m \in \mathcal{M}$; otherwise, the cloud server aggregates the model parameters from edge servers and obtains the evaluating loss $F_c(\boldsymbol{w})$ with the global parameters $\boldsymbol{w}$.

3.4 Problem Formulation and Transformation

In this section, we formulate an optimization problem to manage the spectrum and computing resources and then transform the formulated problem into a POMDP.

3.4.1 Optimization Problem Formulation

As previously described, the FL evaluating loss $F(\boldsymbol{w})$ is used to optimize the performance in the FL process. Thus, to minimize $F(\boldsymbol{w})$ over the aggregated model parameters and IGWs' private datasets while satisfying the FL epoch delay, time budget, and IGWs' long-term energy consumption requirements, we formulate the problem as follows:

$$\mathbf{P}_0: \min_{\boldsymbol{o},\boldsymbol{\xi}^{\downarrow},\boldsymbol{\xi}^{\uparrow},\boldsymbol{\eta}} F(\boldsymbol{w})$$

$$\begin{aligned}
\text{s.t. } & E_{m,n,t} \leq E_{m,n}^{\max}, \forall m \in \mathcal{M}, n \in \mathcal{N}_m, t \in \mathcal{R}, & (3.17a)\\
& T_{m,n,t} \leq T_{m,n}^{\max}, \forall m \in \mathcal{M}, n \in \mathcal{N}_m, t \in \mathcal{R}, & (3.17b)\\
& T_t \leq T^{\max}, \forall t \in \mathcal{R}, & (3.17c)\\
& o_{m,n,t} \in \{0, 1\}, \forall m \in \mathcal{M}, n \in \mathcal{N}_m, t \in \mathcal{R}, & (3.17d)\\
& \sum_{m\in\mathcal{M},n\in\mathcal{N}_m} o_{m,n,t}\xi_{m,n,t}^{\downarrow} = 1, 0 \leq \xi_{m,n,t}^{\downarrow} \leq 1, & (3.17e)\\
& \sum_{m\in\mathcal{M},n\in\mathcal{N}_m} o_{m,n,t}\xi_{m,n,t}^{\uparrow} = 1, 0 \leq \xi_{m,n,t}^{\uparrow} \leq 1, & (3.17f)\\
& 0 \leq \eta_{m,n,t} \leq 1, \forall m \in \mathcal{M}, n \in \mathcal{N}_m, t \in \mathcal{R}, & (3.17g)
\end{aligned}$$

where $\boldsymbol{o} = \{o_{m,n,t}\}_{\forall m\in\mathcal{M},n\in\mathcal{N}_m}$, $\boldsymbol{\xi}^{\downarrow} = \{\xi_{m,n,t}^{\downarrow}\}_{\forall m\in\mathcal{M},n\in\mathcal{N}_m}$, $\boldsymbol{\xi}^{\uparrow} = \{\xi_{m,n,t}^{\uparrow}\}_{\forall m\in\mathcal{M},n\in\mathcal{N}_m}$, and $\boldsymbol{\eta} = \{\eta_{m,n,t}\}_{\forall m\in\mathcal{M},n\in\mathcal{N}_m}$. With these decisions, the model parameters $\boldsymbol{w}$ are optimized in the local model training stage via optimization methods, such as the stochastic gradient decent method [53]. Constraints (3.17a), (3.17b), and (3.17c) ensure the cumulative energy consumption and time, and FL epoch delay can be restricted within $E_{m,n}^{\max}$, $T_{m,n}^{\max}$, and $T^{\max}$, respectively.

Note that although a synchronous mechanism is adopted to aggregate the distributed model parameters, the cumulative delay for each IGW is different since the IGWs would go to sleep after the stage of parameter distribution, local training, or parameter uploading. Thus, the practical service delay in each epoch is different during the FL operation, and the constraint (3.17b) ensures the cumulative delay for each IGW can be restricted within $T_{m,n}^{\max}$. In addition, inappropriate resource allocation decisions may lead to a large delay for parameter distribution, local training, and parameter uploading, and therefore reducing the effectiveness of the FL operation. Thus, the constraint (3.17c) is adopted to ensure the epoch delay can be restricted within $T^{\max}$. Constraint (3.17d) indicates whether the corresponding IGW is selected to participate in FL epoch t. Constraints (3.17e)–(3.17g) guarantee the available bandwidth $B^{\downarrow}$ and $B^{\uparrow}$, and on-device computing resource f can be judiciously allocated.

Through solving the formulated loss minimization problem, FL performance can be efficiently optimized in the considered distributed IIoT networks with satisfied multiple requirements. However, there are some challenges in solving the problem.

- The objective function of problem $\mathbf{P}_0$ is implicit and constraint (3.17a) is temporally correlated, which prevent traditional optimization methods to solve the problem. Due to the strict delay requirement of FL epochs, problem $\mathbf{P}_0$ should be solved with a low latency.
- Problem $\mathbf{P}_0$ is a mix-integer nonlinear optimization problem, which contains both discrete and continuous decision variables. Also, the considered decision variables are correlated with each other, which further complicates decision making process.

The device selection and resource allocation aim at minimizing the long-term evaluating loss, which can be reformulated as a Markov decision process. To obtain the optimal solution, deep RL is an effective approach endowed with strong representation capabilities, which can effectively solve the resource optimization problem with mixed decision variables and temporally correlated constraints [54, 55]. However, single-agent based deep RL algorithms generally require a global controller to collect all the state information from distributed IGWs and observe the entire network environment, which incurs backbone signaling overhead. To address the issue, a deep multi-agent RL algorithm should be adopted, in which each edge controller acts as an agent to cooperatively learn a dynamic resource management policy and solve the problem.

3.4.2 *Problem Transformation with DRL*

To optimize FL in the distributed IIoT networks, we first transform the formulated problem $\mathbf{P}_0$ into a POMDP for M agents [56]. In the considered distributed scenario, the edge controllers are modeled as the *edge agents*, respectively. Each edge agent can observe the state information from their individual environment. For the transformed POMDP, the edge agents can capture the dynamics, i.e., the remaining energy resource of IGWs, from the environment to learn their optimal policies $\{\boldsymbol{\pi}^*_{\theta_m}\}_{\forall m \in \mathcal{M}}$ parameterized by $\boldsymbol{\theta}_m$. The policies can accurately map a set of states $\{\boldsymbol{s}_{m,t}\}_{\forall m \in \mathcal{M}}$ to a set of actions $\{\boldsymbol{a}_{m,t}\}_{\forall m \in \mathcal{M}}$, which is evaluated by the reward function $\{\boldsymbol{r}_{m,t}\}_{\forall m \in \mathcal{M}}$. The state, action, and reward can be defined as follows.

(1) **State**: In FL epoch t, each edge agent collects the state information from their individual environment, which indicates the remaining FL time and energy of IGWs. Namely,

$$\boldsymbol{s}_{m,t} = \{T^r_{m,n,t}, E^r_{m,n,t}\}, \forall m \in \mathcal{M}, n \in \mathcal{N}_m, \tag{3.18}$$

where $T^r_{m,n,t} = T^{\max}_{m,n} - T_{m,n,t-1}$ and $E^r_{m,n,t} = E^{\max}_{m,n} - E_{m,n,t-1}$.

(2) **Action**: According to the observed individual environment states, the edge agents cooperatively make decisions to select the IGWs to participate in the current FL epoch, and allocate the downlink/uplink spectrum and on-device computing resources for supporting the FL, i.e.,

$$\boldsymbol{a}_{m,t} = \{o_{m,n,t}, \xi^{\downarrow}_{m,n,t}, \xi^{\uparrow}_{m,n,t}, \eta_{m,n,t}\}, \forall m \in \mathcal{M}, n \in \mathcal{N}_m. \tag{3.19}$$

Each action element is reshaped into a range of [0, 1]. Note that only when $o_{m,n,t}$ equals 1, the IGW then can use the allocated spectrum and computing resources to support FL, such that constraints (3.17d)–(3.17g) can be satisfied.

(3) **Reward**: Once the action $\boldsymbol{a}_{m,t}$ of edge agent m has been determined based on the observed individual state $\boldsymbol{s}_{m,t}$. The agent will obtain a reward to evaluate the quality of the action. To minimize the FL evaluating loss, the reward function of the multi-agent learning algorithm then can be defined as

$$\boldsymbol{r}_{m,t} = \begin{cases} -F_e(\boldsymbol{w}_m), & \text{if } \mathrm{mod}(t, t_c) \neq 0, \\ -F_c(\boldsymbol{w}), & \text{if } \mathrm{mod}(t, t_c) = 0, \\ -U, & \text{if no aggregation,} \end{cases} \tag{3.20}$$

where $F_e(\boldsymbol{w}_m)$ and $F_c(\boldsymbol{w})$ are the FL evaluating losses of DNN models after the edge and cloud aggregations, respectively. Considering the agents are cooperatively performing the FL operation, $F_e(\boldsymbol{w}_m)$ and $F_c(\boldsymbol{w})$ are set as the rewards for each agent in these two aggregation cases. Note that $F_e(\boldsymbol{w}_m)$ is obtained by a mean value of $F_m(\boldsymbol{w}_m)$ that is defined in Eq. (3.4). An edge at the edge can obtain the mean loss via the following steps. First, the loss value $F_m(\boldsymbol{w}_m)$ evaluated at the edge servers is transmitted to a cloud server. Second, the mean loss, $F_e(\boldsymbol{w}_m) = \frac{1}{M}\sum_{m\in\mathcal{M}} F_m(\boldsymbol{w}_m)$, is calculated by the cloud server. Third, $F_e(\boldsymbol{w}_m)$ is broadcasted to each edge server, which is used as the reward of the edge agent. In addition, U is a penalty factor, which should take a relatively large value to penalize the decisions that cannot aggregate any model parameters within constraints (3.17a)–(3.17c) in each FL epoch. In particular, the minus before the above three terms is mainly used to transform the original loss minimization problem into a reward maximization problem.

In the POMDP, each agent m aims at maximizing its own cumulative discounted reward with the optimal device selection and resource allocation policy $\boldsymbol{\pi}^*_{\theta_m}$. As such, $\mathbf{P}_0$ can be reformulated as multiple distributed optimization sub-problems, i.e.,

$$\mathbf{P}_{m,1}: \max_{\pi_{\theta_m}} \mathbb{E}\left[\sum_{\delta=0}^{\infty} \gamma^{\delta} r_{m,t+\delta} | \boldsymbol{\pi}_{\theta_m}, \boldsymbol{s}_{m,t}, \boldsymbol{a}_{m,t}\right]$$

$$\text{s.t. } (3.17a) - (3.17g), \tag{3.21a}$$

where $\gamma \in [0, 1]$ is the discount factor. Due to the cooperative relationship among the edge agents, sub-problems $\mathbf{P}_{m,1} (\forall m \in \mathcal{M})$ are correlated with each other.

3.5 Multi-agent SAC-Based Scheme for Efficient FL

Taking into account the correlation among sub-problems $\mathbf{P}_{m,1} (\forall m \in \mathcal{M})$, which should be jointly solved to achieve the same goal, i.e., reward maximization. Moreover, due to the mix-integer decision variables, instead of value-based RL algorithms, we adopt a policy gradient-based algorithm, i.e., soft actor-critic (SAC) [57], for each agent to address its POMDP with more robust learning performance.

In this section, we propose the RoF scheme, based on a multi-agent SAC (MASAC)-based resource management algorithm and a device refinement subroutine, to cooperatively solve the modeled POMDP for M edge agents. As illustrated in Fig. 3.2, the proposed MASAC-based algorithm that combines the SAC algorithm with a cooperative multi-agent RL framework [58], is presented. In the following, we first introduce the SAC and multi-agent RL framework, and then present the MASAC and the device refinement subroutine in detail, followed by the computational complexity analysis of the MASAC.

3.5.1 *SAC and Multi-agent RL Framework*

SAC Algorithm: As shown in the right part of Fig. 3.2, each edge agent is based on the SAC algorithm. In the SAC, three main components are incorporated to enhance the performance: (1) entropy maximization, (2) "actor-critic" network architecture, and (3) off-policy formulation, which are detailed as follows.

(1) Maximum entropy is considered to effectively enhance the learning stability and exploration, which aims to maximize both the cumulative discounted reward and the expected entropy of the policy simultaneously. With this maximum entropy objective, the stochasticity of the agent's policy can be greatly improved

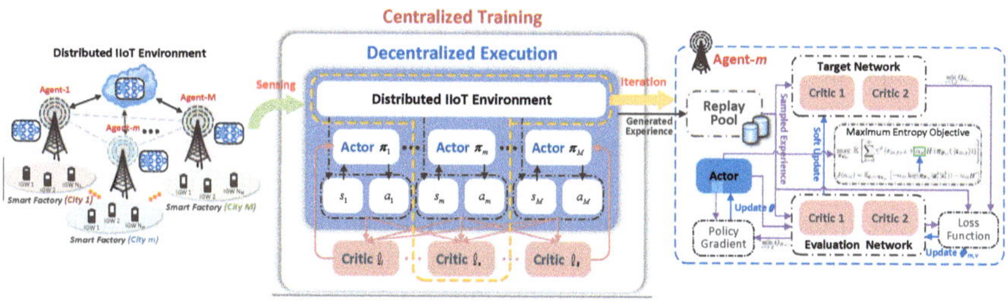

Fig. 3.2 The architecture of the proposed MASAC-based resource management algorithm

while enabling more possible optimal decisions to be explored. As such, the reformulated sub-problems $\mathbf{P}_{m,1} (\forall m \in \mathcal{M})$ can be rewritten as

$$\mathbf{P}_{m,2}: \max_{\pi_{\theta_m}} \mathbb{E}\left[\sum_{\delta=0}^{\infty} \gamma^{\delta}\left(r_{m,t+\delta} + \alpha_m H\left(\pi_{\theta_m}(\cdot|s_{m,t})\right)\right)\right]$$

$$\text{s.t.} \ (3.17a) - (3.17g), \tag{3.22a}$$

where $H(\cdot)$ is the entropy term, and α_m is a weight that controls the relative importance of both terms.

(2) Actor-critic architecture of each edge agent is composed of an actor network, a pair of evaluation critic networks, and a pair of target critic networks. The actor makes decisions $\boldsymbol{a}_{m,t}$ with its own policy π_{θ_m}. The critics calculate a pair of Q-values for evaluating the policy, respectively. As shown in Fig. 3.3, the actor and the critics are endowed with a four-layer fully-connected neural network (FCNN), respectively. Some tricks are used in both networks to improve the effectiveness of the algorithm. For the actor network, the input (i.e., individual states of each agent) should be normalized to avoid overfitting. In the output layer, decision elements should also be addressed via discretization or normalization methods, to satisfy the mixed decision variable constraints. For the critic networks, the input (i.e., action-state pairs) should also be normalized to contribute to better Q-value estimation. Note that the target critics have the same structures as the evaluation critics.

(3) The off-policy formulation is adopted to improve the sample efficiency, which is based on an experience replay technique. Specifically, an experience replay pool $\mathcal{K}_r$ endowed with a certain memory capacity is deployed. In the sampling stage, the experiences of all the agents are stored into the $\mathcal{K}_r$. In the training stage, the agents randomly sample a batch of experiences from $\mathcal{K}_r$ to train the network parameters in an off-policy manner, such that the sampled experiences can be effectively utilized to achieve convergence.

Multi-agent RL Framework: According to the left part of Fig. 3.2, the edge agents operate at the corresponding edge controllers, respectively. Each agent can only observe the local information from its individual environment, and the decisions made by the agent are insensible to the others.

We adopt the mechanism of centralized training while decentralized execution [58], as illustrated in the central part of Fig. 3.2. The mechanism allows the policies to utilize additional information to simplify training process, on condition that this information is not utilized at execution stage. In the centralized training stage, except for the local information, additional information, including the states and actions of all the agents, is also available to each other. Specifically, the actor network of the edge agent captures the environment dynamics only from its own observed local state information, and then makes decisions for the individual environment. The

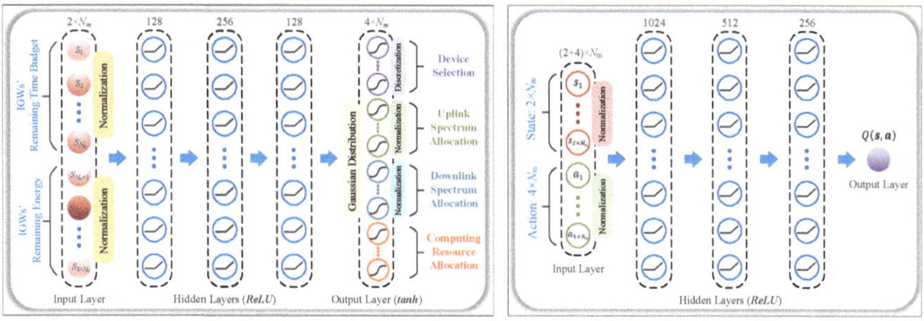

(a) Actor network (b) Critic network

Fig. 3.3 Actor and critic network structures of MASAC

Algorithm 2: MASAC-based algorithm for the JDSRA

Input: Remaining energy and time of all the IGWs $\{T^r_{m,n,t}, E^r_{m,n,t}\}_{\forall m \in \mathcal{M}, n \in \mathcal{N}_m}$;

Output: JDSRA decision $\boldsymbol{a}_t = \{\boldsymbol{o}, \boldsymbol{\xi}^{\downarrow}, \boldsymbol{\xi}^{\uparrow}, \boldsymbol{\eta}\}$;

Initialize actor, evaluation critic, and target critic networks with weights $\boldsymbol{\theta}_m$, $\boldsymbol{\phi}_{m,1}$, $\boldsymbol{\phi}_{m,2}$, $\boldsymbol{\phi}'_{m,1}$, and $\boldsymbol{\phi}'_{m,2}$ for each agent $m \in \mathcal{M}$, and experience replay pool $\mathcal{K}_r$;

for each *episode* **do**

  Receive initial state $\boldsymbol{s}_1 = \{\boldsymbol{s}_{m,1}\}_{\forall m \in \mathcal{M}}$;

  for *FL epoch* $t \in \mathcal{R}$ **do**

    for each *agent* **do**

      ▷ Experience sampling

      Obtain decision $\boldsymbol{a}_{m,t}$ with policy $\pi_{\boldsymbol{\theta}_m}(\boldsymbol{s}_{m,t})$;

      Execute decision $\boldsymbol{a}_{m,t}$, and obtain reward $\boldsymbol{r}_{m,t}$ by Algorithm 1 and next state $\boldsymbol{s}'_{m,t}$;

      Store $\{\boldsymbol{s}_{m,t}, \boldsymbol{a}_{m,t}, \boldsymbol{r}_{m,t}, \boldsymbol{s}'_{m,t}\}$ into $\mathcal{K}_r$ and replace the oldest experiences in $\mathcal{K}_r$;

      ▷ Parameter updating

      Randomly sample a minibatch of K_b experiences $\{\boldsymbol{s}^k_{m,t}, \boldsymbol{a}^k_{m,t}, \boldsymbol{r}^k_{m,t}, \boldsymbol{s}'^k_{m,t}\}$ from $\mathcal{K}_r$;

      Set the target Q-value $y^k_{m,t}$ by Eq. (3.24);

      Update $\boldsymbol{\phi}_{m,1}$ and $\boldsymbol{\phi}_{m,2}$ by minimizing the loss function in Eq. (3.23);

      Update $\boldsymbol{\theta}_m$ via the policy gradient in Eq. (3.28);

      Adjust α_m via the gradient obtained by Eq. (3.29);

    Update target networks for the agents by Eq. (3.30);

    Obtain cooperative decision $\boldsymbol{a}_t = \{\boldsymbol{a}_{m,t}\}_{\forall m \in \mathcal{M}}$.

---

critic network requires the action-state pairs of all the edge agents to generate the Q-values for evaluating the decisions. In the decentralized execution stage, the network parameters of the actor and the critics no longer need to be updated. Thus, only the local state information is required for the actor of each edge agent, the JDSRA decisions then can be obtained independently without being aware of other agents' state information.

Remark: Once well trained, the RL algorithm is deployed to make online decisions for the three-layer FL architecture, which is an inference process. The energy consumption of inference process of the proposed RL algorithm is very low since the adopted neural network has shallow layers, we therefore neglect the energy consumption consumed by the RL algorithm.

3.5.2 *Multi-agent SAC-Based Resource Scheduling Algorithm*

In the following, we illustrate the learning procedure of the MASAC in detail, which is presented in Algorithm 2.

(1) *Initialization (Line 1)*: At the beginning of the centralized training stage, the parameters of the actor and critic networks of the edge agents are initialized, which will be updated in terms of the learning step. In addition, an experience replay pool $\mathcal{K}_r$ is instantiated.
(2) *Experience Sampling (Lines 6–9)*: The experience of the edge agents is denoted by a multi-dimensional tuple of the selected action, state transition, and feedback reward, i.e., $\{\boldsymbol{s}_{m,t}, \boldsymbol{a}_{m,t}, \boldsymbol{r}_{m,t}, \boldsymbol{s}'_{m,t}\}_{\forall m \in \mathcal{M}}$, which is obtained via the following steps. First, the edge agents observe the local state information $\{\boldsymbol{s}_{m,t}\}_{\forall m \in \mathcal{M}}$ from their individual environments, respectively. Second, the actors of the agents independently make the decisions $\{\boldsymbol{a}_{m,t}\}_{\forall m \in \mathcal{M}}$ with their own policy $\{\boldsymbol{\pi}_{\boldsymbol{\theta}_m}\}_{\forall m \in \mathcal{M}}$ according to the local information. Here, $\boldsymbol{a}_{m,t} = \{\boldsymbol{o}_{m,t}, \boldsymbol{\xi}^{\downarrow}_{m,t}, \boldsymbol{\xi}^{\uparrow}_{m,t}, \boldsymbol{\eta}_{m,t}\}$ represents the JDSRA decisions made by edge agent m for optimizing FL under the correlated BS. Then, the agents obtain the reward $\{\boldsymbol{r}_{m,t}\}_{\forall m \in \mathcal{M}}$, and individual environments evolve to the next state $\{\boldsymbol{s}'_{m,t}\}_{\forall m \in \mathcal{M}}$. Finally, the formulated transition tuple $\{\boldsymbol{s}_{m,t}, \boldsymbol{a}_{m,t}, \boldsymbol{r}_{m,t}, \boldsymbol{s}'_{m,t}\}_{\forall m \in \mathcal{M}}$ is stored in the $\mathcal{K}_r$ for algorithm parameter updating.
(3) *FCNN Parameter Updating (Lines 10–17)*: In this stage, the edge agents are coordinated, and the algorithm alternates between collecting experience from all the individual environments with the current agents' policies and updating the FCNNs based on the batch of K_b transitions $\{\boldsymbol{s}^k_{m,t}, \boldsymbol{a}^k_{m,t}, \boldsymbol{r}^k_{m,t}, \boldsymbol{s}'^k_{m,t}\}_{\forall m \in \mathcal{M}}$ sampled from the $\mathcal{K}_r$. For edge agent m, the detailed update procedure is given as follows.

Firstly, the parameters $\boldsymbol{\phi}_{m,v}, \forall v \in \{1, 2\}$ of the evaluation critic networks are independently updated via minimizing the loss function $\mathcal{L}(\boldsymbol{\phi}_{m,v})$, which is given by

$$\mathcal{L}(\boldsymbol{\phi}_{m,v}) = \frac{1}{2K_b} \sum_{k=1}^{K_b} \left(\min_{v=1,2} Q_{\boldsymbol{\phi}_{m,v}} (\boldsymbol{s}^k_t, \boldsymbol{a}^k_{1,t}, \ldots, \boldsymbol{a}^k_{M,t}) - y^k_{m,t} \right)^2, \tag{3.23}$$

where

$$\begin{aligned} y^k_{m,t} = r^k_{m,t} &+ \gamma (\min_{v=1,2} Q_{\boldsymbol{\phi}'_{m,v}} (\boldsymbol{s}'^k_t, \boldsymbol{a}'^k_{1,t}, \ldots, \boldsymbol{a}'^k_{M,t}) \\ &- \alpha_m \log(\boldsymbol{\pi}_{\boldsymbol{\theta}_m}(\boldsymbol{a}'^k_{1,t}, \ldots, \boldsymbol{a}'^k_{M,t} | \boldsymbol{s}'^k_t))) \end{aligned} \tag{3.24}$$

denotes the target Q-value. Here, $s_t^k = \{s_{m,t}^k\}_{\forall m \in \mathcal{M}}$ and $s_t^{\prime k} = \{s_{m,t}^{\prime k}\}_{\forall m \in \mathcal{M}}$ are the current and the next state sets of all the agents, $\log(\cdot)$ is the function to return the actions' entropy value, and $Q_{\boldsymbol{\phi}_{m,v}}(\cdot)$ and $Q_{\boldsymbol{\phi}'_{m,v}}(\cdot)$ are the state-action Q-values calculated by the evaluation and target critic networks. Note that two critics with parameters $\boldsymbol{\phi}_{m,v}$ and $\boldsymbol{\phi}'_{m,v}$, $\forall v \in \{1, 2\}$ are deployed in both the evaluation and target networks, respectively, to alleviate positive bias in policy improvement. More importantly, only the minimum of these two Q-values is used for the loss calculation. As such, the stochastic gradient can be obtained for updating the parameter of the critic networks, i.e.,

$$\begin{aligned}
\nabla_{\boldsymbol{\phi}_{m,v}} \mathcal{L}(\boldsymbol{\phi}_m) = \nabla_{\boldsymbol{\phi}_{m,v}} & Q_{\boldsymbol{\phi}_{m,v}}(s_t^k, \boldsymbol{a}_{1,t}^k, \ldots, \boldsymbol{a}_{M,t}^k) \\
& \cdot (Q_{\boldsymbol{\phi}_{m,v}}(s_t^k, \boldsymbol{a}_{1,t}^k, \ldots, \boldsymbol{a}_{M,t}^k) \\
& - (r_{m,t}^k + \gamma(Q_{\boldsymbol{\phi}'_{m,v}}(s_t^{\prime k}, \boldsymbol{a}_{1,t}^{\prime k}, \ldots, \boldsymbol{a}_{M,t}^{\prime k}) \\
& - \alpha_m \log(\boldsymbol{\pi}_{\boldsymbol{\theta}_m}(\boldsymbol{a}_{1,t}^{\prime k}, \ldots, \boldsymbol{a}_{M,t}^{\prime k} | s_t^{\prime k}))))).
\end{aligned} \tag{3.25}$$

Secondly, the actor network is updated via policy gradient methods, and the objective function is defined as

$$\begin{aligned}
J(\boldsymbol{\theta}_m) = -\mathbb{E}_{s' \sim \mathcal{K}_b}[\mathbb{E}_{\boldsymbol{a}' \sim \boldsymbol{\pi}_{\boldsymbol{\theta}_m}}[& \min_{v=1,2} Q_{\boldsymbol{\phi}'_{m,v}}(s_t^{\prime k}, \boldsymbol{a}_{1,t}^{\prime k}, \ldots, \boldsymbol{a}_{M,t}^{\prime k}) \\
& - \alpha_m \log(\boldsymbol{\pi}_{\boldsymbol{\theta}_m}(\boldsymbol{a}_{1,t}^{\prime k}, \ldots, \boldsymbol{a}_{M,t}^{\prime k} | s_t^{\prime k}))]].
\end{aligned} \tag{3.26}$$

Here, the policy is reparameterized via an FCNN denoted by $g_{\theta_m}(\epsilon_t; s_{m,t})$, in which an input noise ϵ_t is added to obtain a lower variance estimation. Hence, Eq. (3.26) can be rewritten as

$$\begin{aligned}
J(\boldsymbol{\theta}_m) = -\mathbb{E}_{s' \sim \mathcal{K}_b, \epsilon_t \sim \boldsymbol{\pi}_{\boldsymbol{\theta}_m}}[& \min_{v=1,2} Q_{\boldsymbol{\phi}'_{m,v}}(s_t^{\prime k}, g_{\theta_m}(\epsilon_t; s_t^k)) \\
& - \alpha_m \log(\boldsymbol{\pi}_{\boldsymbol{\theta}_m}(g_{\theta_m}(\epsilon_t; s_t^k) | s_t^{\prime k}))],
\end{aligned} \tag{3.27}$$

where ϵ_t is sampled from a Gaussian distribution. Thus, the gradient of the policy can be calculated, i.e.,

$$\begin{aligned}
\nabla_{\theta_m} J(\boldsymbol{\theta}_m) = \nabla_{\theta_m} & \alpha_m \log(\boldsymbol{\pi}_{\boldsymbol{\theta}_m}(\boldsymbol{a}_{1,t}^k, \ldots, \boldsymbol{a}_{M,t}^k | s_t^k)) \\
& + (\nabla_{\boldsymbol{a}_{m,t}} \alpha_m \log(\boldsymbol{\pi}_{\boldsymbol{\theta}_m}(\boldsymbol{a}_{1,t}^k, \ldots, \boldsymbol{a}_{M,t}^k | s_t^k)) \\
& - \nabla_{\boldsymbol{a}_{m,t}} Q(s_t^k, \boldsymbol{a}_{1,t}^k, \ldots, \boldsymbol{a}_{M,t}^k)) \nabla_{\theta_m} g_{\theta_m}(\epsilon_t; s_t^k).
\end{aligned} \tag{3.28}$$

Thirdly, instead of choosing the weight α_m manually, an FCNN is adopted to automate the weight set for the maximum entropy objective. The gradients of α_m can be computed by the following objective, i.e.,

$$J(\alpha_m) = \mathbb{E}_{\boldsymbol{a}_t \sim \boldsymbol{\pi}_{\boldsymbol{\theta}_m}}\left[-\alpha_m \log(\boldsymbol{\pi}_{\boldsymbol{\theta}_m}(\boldsymbol{a}_t^k | s_t^k)) - \alpha_m H'\right], \tag{3.29}$$

Algorithm 3: Device refinement subroutine

Input: Decision $\{\boldsymbol{o}, \boldsymbol{\xi}^{\downarrow}, \boldsymbol{\xi}^{\uparrow}, \boldsymbol{\eta}\}$ obtained by Algorithm 2;
Output: Refined participating device set $\mathcal{N}_m^{\star}$;
Select an initial set of $\mathcal{N}_m, \forall m \in \mathcal{M}$ IGWs for each edge server with device selection decision $\boldsymbol{o}$;
Allocate downlink $\boldsymbol{\xi}^{\downarrow}$ and uplink $\boldsymbol{\xi}^{\uparrow}$ spectrum and on-device computing $\boldsymbol{\eta}$ resources for the selected IGWs;
Calculate the remaining energy $E^r_{m,n,t}$ and time budget $T^r_{m,n,t}$, and FL epoch delay T_t;
Set $\mathcal{N}_m^{\star} = \varnothing$;
for each $BS\ m \in \mathcal{M}$ **do**
    for each $IGW\ n \in \mathcal{N}_m$ **do**
        if $E^r_{m,n,t} \le E^{max}_{m,n}, T^r_{m,n,t} \le T^{max}_{m,n}, and\ T_t \le T^{max}$ **then**
            $\mathcal{N}_m^{\star} \leftarrow \mathcal{N}_m^{\star} \cup \{(m, n)\}$.

where $\boldsymbol{a}_t^k = \{\boldsymbol{a}_{m,t}^k\}_{\forall m \in \mathcal{M}}$ is the decision set of all the agents, and H' is the value of target entropy.

Finally, to stabilize the learning process, the parameters $\boldsymbol{\phi}'_{m,v}$ of target critic networks are updated from the evaluation critics' parameters $\boldsymbol{\phi}_{m,v}$ via a soft-updating method, i.e.,

$$\boldsymbol{\phi}'_{m,v} = \tau\boldsymbol{\phi}_{m,v} + (1-\tau)\boldsymbol{\phi}'_{m,v}, \forall v \in \{1, 2\}, \tag{3.30}$$

where $\tau \in (0, 1)$ is the update factor.

Device Refinement Subroutine In each FL epoch, Algorithm 2 generates the JDSRA decisions to support the FL operation. Based on these decisions, a refinement subroutine is adopted to obtain a refined participating device set that can complete the FL with satisfied energy and delay requirements, thereby accelerating FL convergence. The subroutine is operated as the following steps.

(1) *Remaining On-Device Resource Calculation (Lines 1–4)*: After obtaining the JDSRA decisions, the edge controllers can select an initial set of $\mathcal{N}_m, \forall m \in \mathcal{M}$ IGWs to participate in the FL with the allocated spectrum and computing resources. And then, the remaining on-device energy and IGWs' time budget can be calculated, which are consumed by the FL parameter transmission and local model training.
(2) *Participating Device Set Refinement (Lines 5–8)*: With the above calculations, the edge controllers can evaluate whether the selected IGWs are able to complete the FL while satisfying the strict requirements. Only when the IGWs can satisfy the above requirement, the system would allow these IGWs to participate in the FL. Thus, a refined set of participating IGWs $\mathcal{N}_m^{\star}, \forall m \in \mathcal{M}$ can be obtained to optimize FL.

With this subroutine, the on-device energy consumption and time budget can be effectively reduced to accelerate convergence in the distributed and resource-limited IIoT networks.

Computation Complexity Analysis As described before, the MASAC algorithm contains M edge agents. Each agent is endowed with one actor network, four critic networks, and one α_m weight networks. Each of them is represented by an FCNN, as shown in Fig. 3.3. Generally, the computational complexity of an FCNN is $K = \mathcal{O}(\sum_x (2I_x - 1)I_{x+1})$ [59], where $x \in [0, L]$ denotes the layer index and I_x represents the neuron number of hidden layers. According to the different input-output structures, there are three cases for the above networks. Specifically,

(1) For actor networks, $I_1 = 2|\mathcal{N}_m|, \forall m \in \mathcal{M}$ is the dimension of the corresponding individual environment states, and $I_L = 4|\mathcal{N}_m|$ is the decision dimension of edge agent m;
(2) For critic networks, $I_1 = (2+4)\sum_m |\mathcal{N}_m|$ is the dimension of total states and decisions. Here, the total states are composed of all the individual states observed by the agents of dimension $2\sum_m |\mathcal{N}_m|$. The decisions include device selection variable, and uplink/downlink spectrum and computing resource allocation variables with dimension $4\sum_m |\mathcal{N}_m|$. Note that the output of the critic network is an estimated Q-value with dimension 1, and thus $I_L = 1$;
(3) For α_m weight networks, the network scale is very small, and thus the computational complexity can be neglected.

In summary, there are $5 \times M$ FCNNs should be considered in the MASAC, thus the total computational complexity of the algorithm is $5 \times M \times K$.

Moreover, the proposed RoF scheme can reduce signaling overhead. During the FL operation, the central controller deployed at the cloud server collects the network state information from IGWs in each epoch. In the single-agent RoF scheme, the state information has to be sent to the central controller. In the RoF scheme, the state information only needs to be sent to the edge controllers that are deployed at edge servers. That is, compared with the single-agent RoF scheme, the RoF scheme can effectively reduce the signaling overhead of reporting all the state information to the cloud server.

3.6 Performance Evaluation and Analysis

In this part, simulations are conducted to evaluate the proposed RoF scheme for optimizing FL in the distributed IIoT networks.

We consider three smart factories operating in different cities, respectively, and 30 IGWs are distributed in each factory that is covered by a BS unless specified. The channel gain in IIoT networks is modeled by $PL(\text{dB}) = -128.1 - 37.6\log_{10}(l)$, where l denotes the transmission distance between the BS and the covered IGWs [60]. We adopt Raspberry Pi 4B as an IGW [61]. The effective capacitance is set to be 2×10^{-28}, and the number of CPU cycles to execute one training sample is set to be

Table 3.1 Simulation parameters [57, 62]

Network para.	Value	Learning para.	Value
$p_{m,n}$, p_m	30, 40 dBm	*Optimizer*	Adam
σ^2	− 104 dBm	*Episode number*	300
l	150 m	*Epoch number*	300
$B^{\uparrow}$, $B^{\downarrow}$	20, 40 MHz	K_r	10,000
f	1.5 GHz	K_b	64
R^c	1 Mbps	γ	0.9
$E_{m,n}^{\max}$	[4500, 5000] J	ϵ	3×10^{-4}
$T_{m,n}^{\max}$	[600, 800] s	τ	0.01
$T^{\max}$	5 s	U	10

20 cycles/bit [50]. For the MASAC algorithm, the actor and the critics of each edge agent are with [128, 256, 128, 120] and [1024, 512, 256, 1] neurons, respectively. Note that each agent employs the same actor and critic network architecture. Other important simulation parameters are listed in Table 3.1.

For the FL, we design a convolutional neural network (CNN)-based model [63], which includes two 5×5 convolutional layers, two fully-connected layers, and a softmax output layer (61,366 total trainable model parameters), to perform classification tasks (i.e., fault diagnostic). In addition, the learning rate and batch size of the CNN model are set to be 2.5×10^{-3} and 64, respectively. The regularization coefficient is set to be 5×10^{-5}. In this simulation, the CNN is trained to perform fault diagnostic task, and thus the loss function can be expressed as

$$f(\boldsymbol{x}_d, y_d; \boldsymbol{w}) = \sum_{z=0}^{Z} p(y_d = z)\mathbb{E}_{\boldsymbol{x}|y_d=z}[\log \hat{y}_d], \tag{3.31}$$

where Z is the number of health condition classes, y_d is the true label, and $\hat{y}_d$ is the predicted label [25].

The data size of model parameters S_g is 240 kB. In addition, the CNN model is trained on the AWE dataset,[2] which is a vibration-based industrial sensor dataset contains six health condition classes of industrial facilities [64]. In our simulations, the number of data samples in the AWE dataset is 23,436, and the ratio between the training set and evaluating set is 9:1. Each sample contains 1024 sampling points. Figure 3.4 shows two raw samples of the AWE dataset, which are collected from the "Normal" and "Fault-1" health conditions, respectively. In the considered scenario, FL is operated in a non-IID setting, in which each IGW only holds a few classes of samples. To achieve this, we sort the samples by the class label, partition them into 180 blocks with the same size, and assign 2 blocks to each IGW [14].

[2] Online available. https://github.com/Intelligent-AWE/DeepHealth.

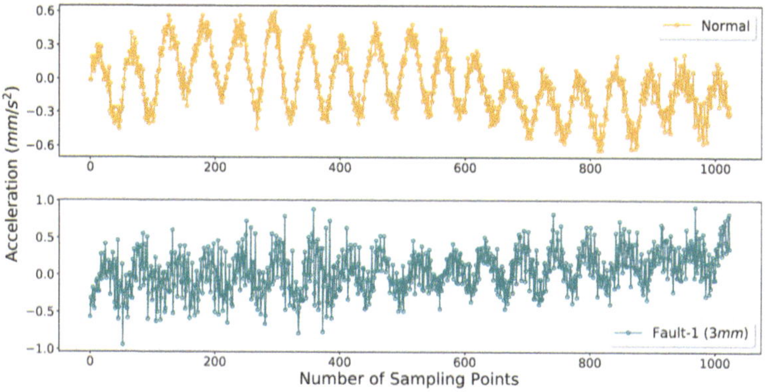

Fig. 3.4 Two classes of samples in AWE dataset. These two samples are collected from rolling bearings of industrial facilities. "Normal" means the bearing stays in a health condition, and "Fault-1" means the bearing operates under a fault condition with 3 mm severity degree

To evaluate the performance of the proposed scheme, the following benchmarks are compared.

- ***Single-Agent RoF (SARoF)***: In this scheme, we assume a central controller is deployed at the cloud server and acts as the agent to centrally select participating IGWs and manage the spectrum and computing resources for the FL. In the SARoF, the hidden layers of the actor network are with [256, 512, 384, 360] neurons, and the critic networks have the same structure as that of the RoF.
- ***The RoF without Edge Aggregation (RoF w.o.)***: This scheme is designed for a two-layer FL aggregation architecture, in which the FL model parameters are only aggregated by the cloud server in each FL epoch.
- ***Multi-Agent DDPG (MADDPG)***: In this scheme, the hidden layers of the actor networks and the critic networks are with [180, 300] and [450, 100] neurons, respectively. Multiple learning agents are deployed at the edge servers to cooperatively make the JDSRA decisions via a DDPG algorithm.
- ***Random Probabilistic Configuration***: In this scheme, the random policy is adopted, in which the central controller randomly selects the IGWs and allocates the spectrum and computing resources. All available actions are selected with an equal probability.

We evaluate the convergence performance of the three-layer collaborative FL for the CNN-based fault diagnostic model. During the FL operation, the cloud aggregation is performed every t_c FL epochs. As shown in Fig. 3.5, the CNN model trained on the AWE dataset can achieve a low evaluating loss and a high accuracy over the considered three-layer collaborative architecture within 300 FL epochs. In particular, lower FL evaluating loss and higher accuracy can be reached if the cloud aggregation frequency is set properly. It can be seen that the lowest evaluating loss and the highest accuracy are respectively obtained when the cloud aggregation frequency t_c is set to be 10. Moreover, due to the non-IID setting and the limited on-device energy,

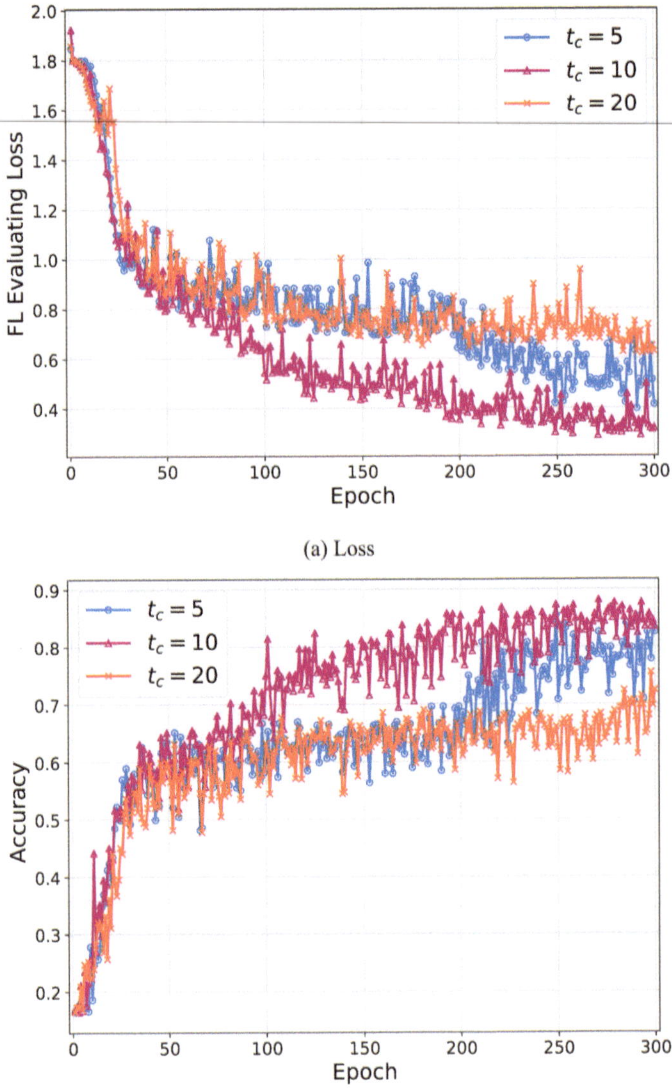

(a) Loss

(b) Accuracy

Fig. 3.5 FL evaluating loss and accuracy with respect to different cloud aggregation frequencies

the FL may not benefit from a large cloud aggregation frequency. This is because on-device energy might be exhausted before cloud aggregation, such that the model parameters trained on these IGWs cannot be globally aggregated at the cloud server, which slows down global model convergence.

As shown in Fig. 3.6, the convergence performance of the CNN-based model in terms of learning rates is evaluated. It can be seen that the lowest evaluating loss and the highest accuracy are respectively achieved when the learning rate is set to be 2.5×10^{-3}. The result indicates that a larger learning rate may not contribute to achieve a better performance. Also, a smaller value of learning rate can lead to a stable FL process but is not good for achieving a lower loss, and thus may consume more spectrum and computing resources to achieve convergence.

To demonstrate the scalability of the proposed solution, we evaluates the performance of the RoF scheme in terms of different numbers of IGWs dispersed in each factory. As shown in Fig. 3.7, the CNN model can achieve a low evaluating loss and a high accuracy within 300 FL epochs. In particular, the achieved evaluating loss and accuracy do not degrade much as the increase of the number of IGWs. The result indicates that the presented RoF scheme can be extended to dense networks with a large number of IGWs. However, with the increasing of the number of IGWs, the decision dimension of the RoF scheme increases, which increases the difficulty for the optimal decision making.

To demonstrate the effectiveness of the learning-based RoF scheme, we evaluate the average accuracy and evaluating loss of different schemes with respect to different amounts of downlink spectrum resources, as shown in Fig. 3.8. As expected, the RoF achieves the highest average accuracy and lowest average evaluating loss among all the schemes. Specifically, the accuracies incurred by the learning-based schemes (i.e., RoF, SARoF, and RoF w.o.) increase with the growth of the amounts of downlink spectrum resources. On the contrary, the evaluating loss achieved by the learning-based schemes decreases with the decrease of the amounts of downlink spectrum resources. The accuracy achieved by the RoF is approximately 1.8% higher than that by the SARoF scheme. The result indicates that the RoF can achieve a higher resource utilization and is more scalable compared with other schemes. In addition, the RoF can increase the average accuracy by 14.6%, as compared with the RoF without edge aggregation. This is because two-layer (i.e., device-cloud) FL architecture leads to higher parameter transmission delay, which prevents the cloud server from aggregating more model parameters with the FL strict epoch delay constraint and thus against FL performance improvement.

As shown in Fig. 3.9, the average accuracy of different algorithms in terms of on-device computing capabilities is evaluated. The result shows that all learning-based schemes, i.e., RoF, SARoF, MADDPG, and RoF w.o., can achieve higher accuracy than that of RPC benchmark. This is because that learning-based schemes make the FL models converge to a high accuracy level via judicious computing resource allocation. In addition, with the increase of the available computing capabilities, the average accuracies achieved by the RoF, SARoF, MADDPG, and RoF w.o. schemes increase. The underlying reason is that the delay of local model training can be greatly reduced with more available computing resources, thereby ensuring the FL can be completed within strict delay requirements. Particularly, the proposed multi-agent empowered RoF scheme can achieve the highest average accuracy than the benchmarks. When the amount of computing capability is 1.5 GHz, the RoF can

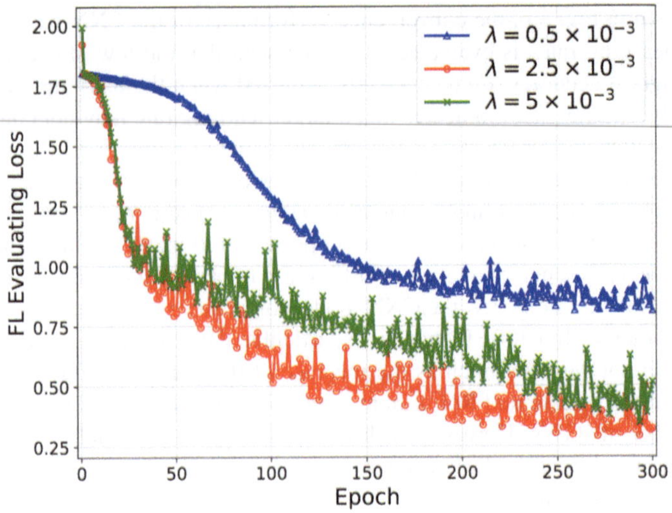

(a) Loss

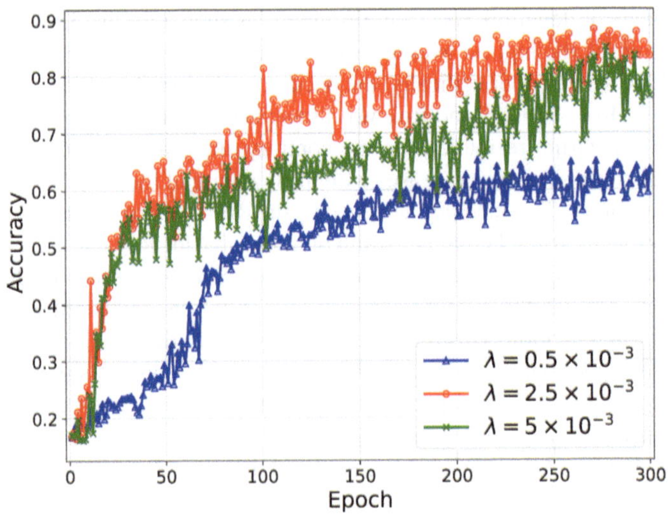

(b) Accuracy

Fig. 3.6 FL evaluating loss and accuracy with respect to different learning rates

increase the average accuracy by 1.81%, 5.47%, and 14.56% compared with the SARoF, MADDPG, and RoF w.o. benchmarks, respectively.

Efficient uplink spectrum resource utilization ensures the parameter aggregation can be completed within the strict FL epoch delay requirement. As shown in Fig. 3.10, the impact of uplink spectrum resources on the aggregated model number of different

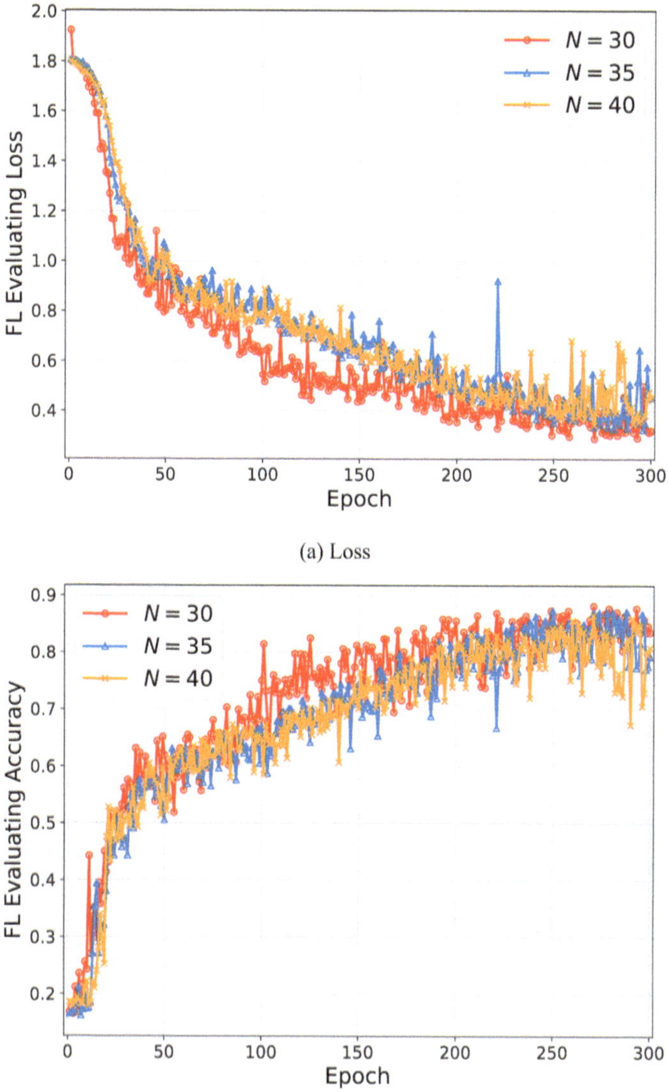

(a) Loss

(b) Accuracy

Fig. 3.7 FL evaluating loss and accuracy with respect to different device (i.e., the IGW) number

schemes is evaluated. Intuitively, more IGWs' model parameters can be aggregated by the edge or cloud servers when there are more available uplink spectrum resources. More importantly, the RoF achieves the maximum average aggregated model number among all the schemes, implying that more aggregated model parameters can be utilized to the optimal global FL model. In particular, when the amount of uplink

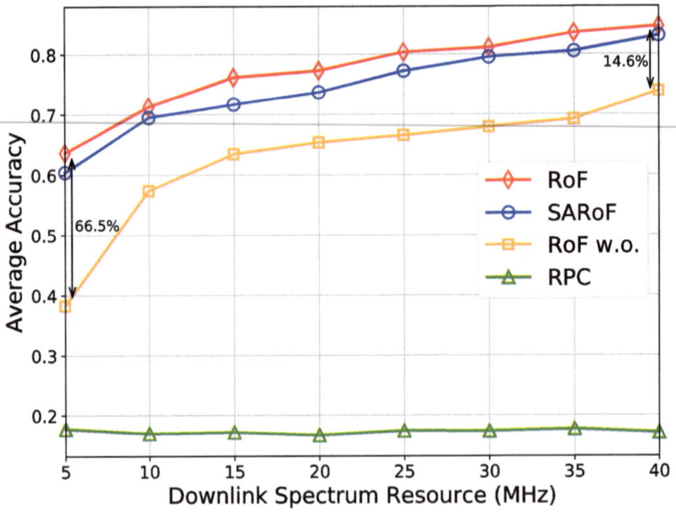

(a) Accuracy

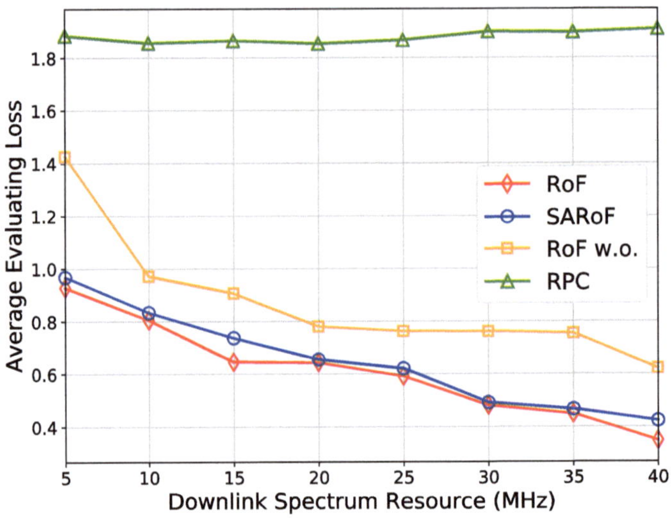

(b) Loss

Fig. 3.8 Average accuracy and evaluating loss comparison among different schemes with respect to downlink spectrum resource

spectrum resource is 20 MHz, the RoF increases the average number of aggregated model by 30.8% compared with the SARoF scheme. This is because the RoF can better allocate the uplink spectrum resources, thereby enabling the parameter uploading to be completed with an acceptable transmission delay. With more model parameters

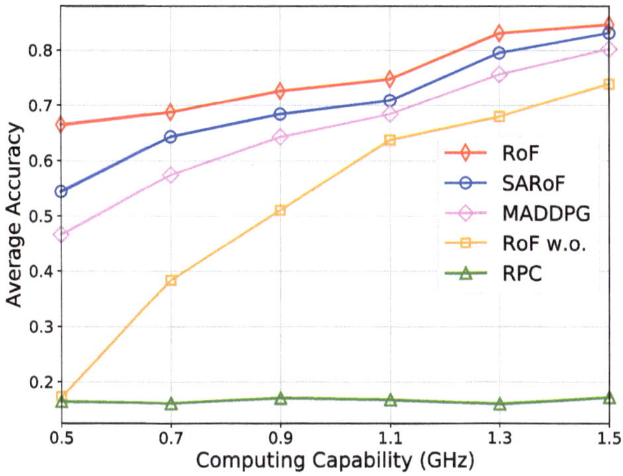

Fig. 3.9 Average accuracy comparison among different schemes with respect to computing capability

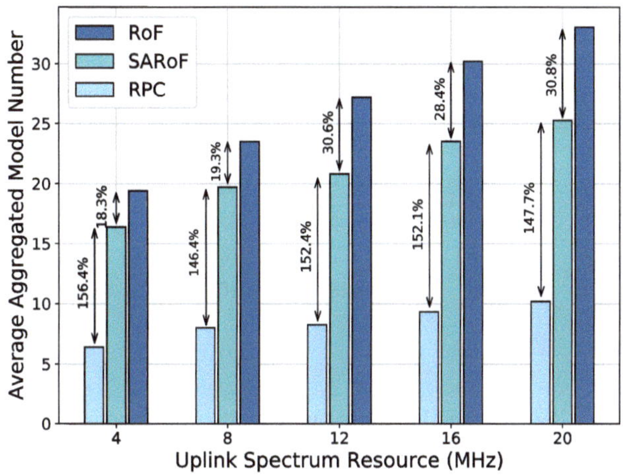

Fig. 3.10 Average aggregated model number comparison among different schemes with respect to uplink spectrum resource

aggregated from the distributed IGWs, the FL models can achieve a higher accuracy level, as the red curve shown in Fig. 3.8. The result indicates that implementing FL over more usable data can effectively optimize the FL performance.

As shown in Fig. 3.11, the average energy consumption and time of different algorithms in terms of uplink spectrum resource are presented. The RoF can effectively reduce the average energy consumption and time as compared to the SARoF scheme. This is because the proposed RoF scheme can better utilize uplink spectrum resources to enable FL models to be converged within energy consumption and time

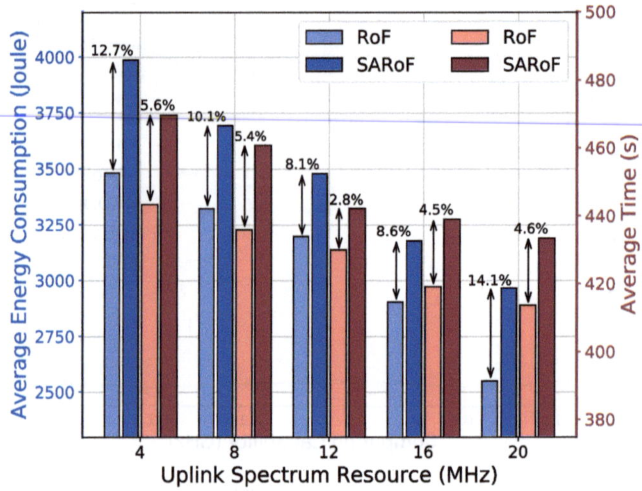

Fig. 3.11 Average energy consumption and time of different schemes with respect to uplink spectrum resource

Table 3.2 Performance comparison of different schemes

	Average accuracy (%)	Average loss	Average participating num.	Average aggregated num.	Average energy (J)	Average time (s)
RoF	84.5	0.3451	53.8	33.1	2550.3	413.8
SARoF	83.1	0.4209	36.5	25.3	2967.2	433.5
RoF w.o.	73.8	0.62	36.2	10.9	2539.3	190.5
RPC	17.0	1.81	44.8	10.2	4800	11.5

requirements in a decentralized resource management manner. In addition, both the energy consumption and time decrease with the increase of the amounts of available spectrum resource. Specifically, when the amount of uplink spectrum resource is 20 MHz, the RoF reduces the average energy consumption and time within the FL process by approximately 14.1% and 4.6%, respectively. The reason is that sufficient spectrum resource enables the model parameters to be aggregated at the edge or cloud with a shorter parameter uploading time, thereby consuming less on-device energy for parameter uploading. In Fig. 3.12, the cumulative distribution functions of the energy consumption and time of different schemes are compared, which shows that the RoF scheme can better guarantee the energy consumption and time requirements.

As listed in Table 3.2, a comprehensive comparison is presented to demonstrate the performance of different schemes. Overall, three important observations can be obtained from the simulation results. Firstly, the RoF achieves the optimal performance over all the metrics given multiple constrained resources. Specifically, the average accuracy of the RoF is about 84.5%, which is higher than that of the SARoF

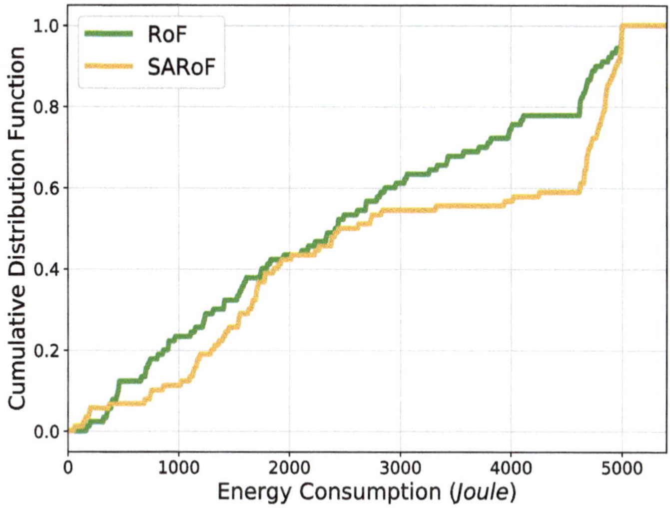

(a) Energy

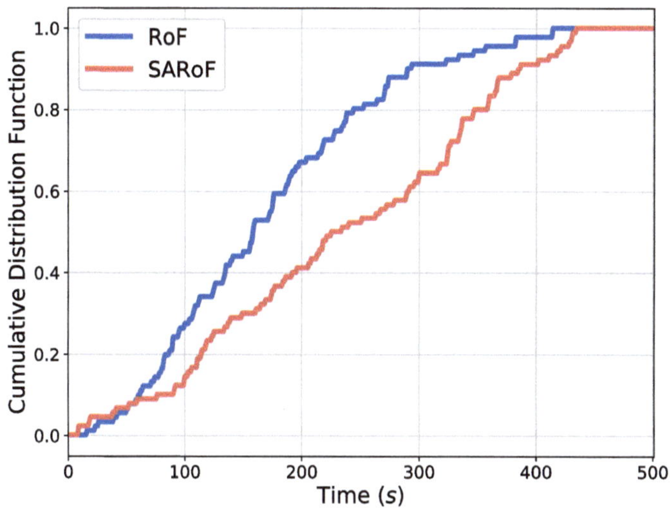

(b) Time

Fig. 3.12 Cumulative distribution functions of energy consumption and time during the FL process

(83.1%), RoF w.o. (73.8%), and RPC (17.0%). The average loss of the RoF is about 0.35, which is lower than that of the SARoF (0.42), RoF w.o. (0.62), and RPC (1.81). Secondly, aiming at achieving fast FL convergence, the RoF is inclined to select more IGWs to participate in the FL. However, it is interesting to note that, the ratio between the number of aggregated model and the number of participating IGW for RoF is

61.4%, which is lower than that of SARoF (69.3%). Constrained computing and spectrum resources have significant effects on the local model training and parameter transmission, which should be better utilized to support the long-term FL process, thereby providing reliable distributed learning service. Thirdly, although more IGWs are selected by the RoF during the FL process, they consumes less energy and time than the SARoF. The result implies that the RoF can allocate the cherished resources in a more judicious way, which endows more persistence to the system. The above simulation results indicate that the considered trade-offs should be better balanced to facilitate efficient FL in the distributed IIoT networks.

Based on the above observations, we can claim that the proposed RoF scheme can effectively improve the FL performance in the distributed and resource-limited IIoT networks.

3.7 Summary

In this chatper, we have investigated the FL performance optimization problem in the distributed IIoT networks. To solve the problem, we have proposed the RoF scheme, based on deep multi-agent reinforcement learning, to make the optimal device selection and resource allocation decisions in an online manner. Simulation results based on real-world dataset demonstrate that the proposed scheme can achieve a high FL accuracy, while satisfying on-device energy consumption requirements. The RoF scheme operates in a distributed manner, which can effectively reduce system signaling overhead of collecting network information, especially in resource-constrained networks. For the future work, we will study a data importance-aware device selection scheme to optimize FL in a large-scale IIoT network.

References

1. M. Chen, U. Challita, W. Saad, C. Yin, M. Debbah, Artificial neural networks-based machine learning for wireless networks: a tutorial. IEEE Commun. Surv. Tutor. **21**(4), 3039–3071 (2019)
2. T. Qiu, J. Chi, X. Zhou, Z. Ning, M. Atiquzzaman, D.O. Wu, Edge computing in industrial internet of things: architecture, advances and challenges. IEEE Commun. Surv. Tutor. **22**(4), 2462–2488 (2020)
3. E. Sisinni, A. Saifullah, S. Han, U. Jennehag, M. Gidlund, Industrial internet of things: challenges, opportunities, and directions. IEEE Trans. Ind. Inform. **14**(11), 4724–4734 (2018)
4. W. Saad, M. Bennis, M. Chen, A vision of 6G wireless systems: applications, trends, technologies, and open research problems. IEEE Netw. **34**(3), 134–142 (2019)
5. R.A. Khalil, N. Saeed, M. Masood, Y.M. Fard, M.-S. Alouini, T.Y. Al-Naffouri, Deep learning in the industrial internet of things: potentials, challenges, and emerging applications. IEEE Internet Things J. **8**(14), 11016–11040 (2021)
6. H. Li, G. Hu, J. Li, M. Zhou, Intelligent fault diagnosis for large-scale rotating machines using binarized deep neural networks and random forests. IEEE Trans. Autom. Sci. Eng. **19**(2), 1109–1119 (2021)

7. H. Zhou, C. She, Y. Deng, M. Dohler, A. Nallanathan, Machine learning for massive industrial internet of things. IEEE Wireless Commun. **28**(4), 81–87 (2021)
8. Y. Chang, J. Chen, W. Wu, T. Pan, Z. Zhou, S. He, Intelligent fault quantitative identification for industrial internet of things (IIoT) via a novel deep dual reinforcement learning model accompanied with insufficient samples. IEEE Internet Things J. **9**(20), 19811–19822 (2022)
9. X. Shen, C. Huang, D. Liu, L. Xue, W. Zhuang, R. Sun, B. Ying, Data management for future wireless networks: architecture, privacy preservation, and regulation. IEEE Netw. **35**(1), 8–15 (2021)
10. D. Gündüz, P. Kerret, N. Sidiropoulos, D. Gesbert, C. Murthy, M. Schaar, Machine learning in the air. IEEE J. Sel. Areas Commun. **37**(10), 2184–2199 (2019)
11. K. Huang, G. Zhu, C. You, J. Zhang, Y. Du, D. Liu, Communication, computing, and learning on the edge, in *Proceedings of ICCS*, Chengdu, China (2018)
12. M. Chen, Z. Yang, W. Saad, C. Yin, H. Poor, S. Cui, A joint learning and communications framework for federated learning over wireless networks. IEEE Trans. Wireless Commun. **20**(1), 269–283 (2020)
13. M. Bennis, Federated learning and control at the wireless network edge. GetMobile Mob. Comput. Commun. **24**(3), 9–13 (2021)
14. B. McMahan, E. Moore, D. Ramage, S. Hampson, B. Arcas, Communication-efficient learning of deep networks from decentralized data, in *Proceedings of PMLR AIS*, Florida, USA (2017), pp. 1273–1282
15. M. Chen, H. Poor, W. Saad, S. Cui, Convergence time minimization of federated learning over wireless networks, in *Proceedings of IEEE ICC*, Dublin, Ireland (2020)
16. M. Wadu, S. Samarakoon, M. Bennis, Joint client scheduling and resource allocation under channel uncertainty in federated learning. IEEE Trans. Commun. **69**(9), 5962–5974 (2021)
17. S. Wang, T. Tuor, T. Salonidis, K. Leung, C. Makaya, T. He, K. Chan, Adaptive federated learning in resource constrained edge computing systems. IEEE J. Sel. Areas Commun. **37**(6), 1205–1221 (2019)
18. S. Pandey, N. Tran, M. Bennis, Y. Tun, A. Manzoor, C. Hong, A crowdsourcing framework for on-device federated learning. IEEE Trans. Wireless Commun. **19**(5), 3241–3256 (2020)
19. J. Kang, Z. Xiong, D. Niyato, Y. Zou, Y. Zhang, M. Guizani, Reliable federated learning for mobile networks. IEEE Wireless Commun. **27**(2), 72–80 (2020)
20. W. Lim, N. Luong, D. Hoang, Y. Jiao, Y. Liang, Q. Yang, D. Niyato, C. Miao, Federated learning in mobile edge networks: a comprehensive survey. IEEE Commun. Surv. Tutor. **22**(3), 2031–2063 (2020)
21. X. Chen, Z. Zhao, C. Wu, M. Bennis, H. Liu, Y. Ji, H. Zhang, Multi-tenant cross-slice resource orchestration: a deep reinforcement learning approach. IEEE J. Sel. Areas Commun. **37**(10), 2377–2392 (2019)
22. H. Peng, X. Shen, Deep reinforcement learning based resource management for multi-access edge computing in vehicular networks. IEEE Trans. Netw. Sci. Eng. **7**(4), 2416–2428 (2020)
23. L. Chen, J. Xu, Seek common while shelving differences: orchestrating deep neural networks for edge service provisioning. IEEE J. Sel. Areas Commun. **39**(1), 251–264 (2021)
24. K. Yang, T. Jiang, Y. Shi, Z. Ding, Federated learning via over-the-air computation. IEEE Trans. Wireless Commun. **19**(3), 2022–2035 (2020)
25. H. Wang, Z. Kaplan, D. Niu, B. Li, Optimizing federated learning on non-IID data with reinforcement learning, in *Proceedings of IEEE INFOCOM*, Beijing, China (2020), pp. 1698–1707
26. R. Yu, P. Li, Toward resource-efficient federated learning in mobile edge computing. IEEE Netw. **35**(1), 148–155 (2021)
27. Z. Wang, Q. Hu, R. Li, M. Xu, Z. Xiong, Incentive mechanism design for joint resource allocation in blockchain-based federated learning. IEEE Trans. Parallel Distrib. Syst. **34**(5), 1536–1547 (2023)
28. W. Wen, Z. Chen, H.H. Yang, W. Xia, T.Q.S. Quek, Joint scheduling and resource allocation for hierarchical federated edge learning. IEEE Trans. Wireless Commun. **21**(8), 5857–5872 (2022)

29. S. Liu, G. Yu, X. Chen, M. Bennis, Joint user association and resource allocation for wireless hierarchical federated learning with IID and non-IID data. IEEE Trans. Wireless Commun. **21**(10), 7852–7866 (2022)
30. R. Chen, L. Li, K. Xue, C. Zhang, M. Pan, Y. Fang, Energy efficient federated learning over heterogeneous mobile devices via joint design of weight quantization and wireless transmission. IEEE Trans. Mob. Comput. **22**(12), 7451–7465 (2022)
31. X. Chen, G. Zhu, Y. Deng, Y. Fang, Federated learning over multihop wireless networks with in-network aggregation. IEEE Trans. Wireless Commun. **21**(6), 4622–4634 (2022)
32. X. Cao, T. Başar, S. Diggavi, Y.C. Eldar, K.B. Letaief, H.V. Poor, J. Zhang, Communication-efficient distributed learning: an overview. IEEE J. Sel. Areas Commun. **41**(4), 851–873 (2023)
33. Q. Zeng, Y. Du, K. Huang, K. Leung, Energy-efficient radio resource allocation for federated edge learning, in *Proceedings of IEEE ICC*, Dublin, Ireland (2020)
34. L. Liang, H. Ye, G. Li, Spectrum sharing in vehicular networks based on multi-agent reinforcement learning. IEEE J. Sel. Areas Commun. **37**(10), 2282–2292 (2019)
35. W. Wu, N. Chen, C. Zhou, M. Li, X. Shen, W. Zhuang, X. Li, Dynamic RAN slicing for service-oriented vehicular networks via constrained learning. IEEE J. Sel. Areas Commun. **39**(7), 2076–2089 (2021)
36. N. Zhao, H. Wu, F. Yu, L. Wang, W. Zhang, V. Leung, Deep reinforcement learning-based latency minimization in edge intelligence over vehicular networks. IEEE Internet Things J. **9**(2), 1300–1312 (2022)
37. W. Ahsan, W. Yi, Y. Liu, A. Nallanathan, A reliable reinforcement learning for resource allocation in uplink NOMA-URLLC networks. IEEE Trans. Wireless Commun. **21**(8), 5989–6002 (2022)
38. S. Choi, S. Choi, G. Lee, S.-G. Yoon, S. Bahk, Deep reinforcement learning for scalable dynamic bandwidth allocation in RAN slicing with highly mobile users. IEEE Trans. Veh. Technol. **73**(1), 576–590 (2023)
39. Y. Chiang, C.-H. Hsu, G.-H. Chen, H.-Y. Wei, Deep Q-learning-based dynamic network slicing and task offloading in edge network. IEEE Trans. Netw. Serv. Manag. **20**(1), 369–384 (2022)
40. H. Zhang, H. Wang, Y. Li, K. Long, A. Nallanathan, DRL-driven dynamic resource allocation for task-oriented semantic communication. IEEE Trans. Commun. **71**(7), 3992–4004 (2023)
41. J.A. Ansere, E. Gyamfi, V. Sharma, H. Shin, O.A. Dobre, T.Q. Duong, Quantum deep reinforcement learning for dynamic resource allocation in mobile edge computing-based IoT systems. IEEE Trans. Wireless Commun. **23**(6), 6221–6233 (2023)
42. J.A. Ansere, E. Gyamfi, Y. Li, H. Shin, O.A. Dobre, T. Hoang, T.Q. Duong, Optimal computation resource allocation in energy-efficient edge IoT systems with deep reinforcement learning. IEEE Trans. Green Commun. Netw. **7**(4), 2130–2142 (2023)
43. L.T. Hoang, C.T. Nguyen, A.T. Pham, Deep reinforcement learning-based online resource management for UAV-assisted edge computing with dual connectivity. IEEE/ACM Trans. Netw. **31**(6), 2761–2776 (2023)
44. X. Tan, L. Zhou, H. Wang, Y. Sun, H. Zhao, B.-C. Seet, J. Wei, V.C.M. Leung, Cooperative multi-agent reinforcement-learning-based distributed dynamic spectrum access in cognitive radio networks. IEEE Internet Things J. **9**(19), 19477–19488 (2022)
45. A. Omidkar, A. Khalili, H.H. Nguyen, H. Shafiei, Reinforcement-learning-based resource allocation for energy-harvesting-aided D2D communications in IoT networks. IEEE Internet Things J. **9**(17), 16521–16531 (2022)
46. X. Zhou, X. Zheng, X. Cui, J. Shi, W. Liang, Z. Yan, L.T. Yang, S. Shimizu, I. Kevin, K. Wang, Digital twin enhanced federated reinforcement learning with lightweight knowledge distillation in mobile networks. IEEE J. Sel. Areas Commun. **41**(10), 3191–3211 (2023)
47. K. Li, W. Ni, E. Tovar, A. Jamalipour, On-board deep Q-network for UAV-assisted online power transfer and data collection. IEEE Trans. Veh. Technol. **68**(12), 12215–12226 (2019)
48. A. Kaur, K. Kumar, Energy-efficient resource allocation in cognitive radio networks under cooperative multi-agent model-free reinforcement learning schemes. IEEE Trans. Netw. Service Manag. **17**(3), 1337–1348 (2020)

49. W. Zhang, D. Yang, W. Wu, H. Peng, H. Zhang, X. Shen, Spectrum and computing resource management for federated learning in distributed IIoT, in *Proceedings of IEEE ICC Workshop*, Montreal, Canada (2021)
50. L. Liu, J. Zhang, S. Song, K. Letaief, Client-edge-cloud hierarchical federated learning, in *Proceedings of IEEE ICC*, Dublin, Ireland (2020)
51. W. Wu, N. Cheng, N. Zhang, P. Yang, W. Zhuang, X. Shen, Fast mmWave beam alignment via correlated bandit learning. IEEE Trans. Wireless Commun. **18**(12), 5894–5908 (2019)
52. F. Luo, C. Zhang, *Signal Processing for 5G: Algorithms and Implementations* (Wiley, 2016)
53. L. Bottou, Large-scale machine learning with stochastic gradient descent, in *Proceedings of COMPSTAT*, Paris, France (2010), pp. 177–186
54. S. Deng, Z. Xiang, P. Zhao, J. Taheri, H. Gao, J. Yin, A. Zomaya, Dynamical resource allocation in edge for trustable internet-of-things systems: a reinforcement learning method. IEEE Trans. Ind. Inform. **16**(9), 6103–6113 (2020)
55. X. Shen, J. Gao, W. Wu, K. Lyu, M. Li, W. Zhuang, X. Li, J. Rao, AI-assisted network-slicing based next-generation wireless networks. IEEE Open J. Veh. Technol. **1**(1), 45–66 (2020)
56. J. Cui, Y. Liu, A. Nallanathan, Multi-agent reinforcement learning-based resource allocation for UAV networks. IEEE Trans. Wireless Commun. **19**(2), 729–743 (2019)
57. T. Haarnoja, A. Zhou, P. Abbeel, S. Levine, Soft actor-critic: off-policy maximum entropy deep reinforcement learning with a stochastic actor, in *Proceedings of ICML*, Stockholm, Sweden (2018)
58. R. Lowe, Y. Wu, A. Tamar, J. Harb, P. Abbeel, I. Mordatch, Multiagent actor-critic for mixed cooperative-competitive environments, in *Proceedings of NeurIPS*, Long Beach, United States (2017)
59. R. Livni, S. Shalev, O. Shamir, On the computational efficiency of training neural networks, in *Proceedings of NeurIPS*, vol. 27, Montréal, Canada (2014), pp. 855–863
60. G. Zhao, B. Huang, L. Li, X. Zhou, Relay-assisted cross-channel gain estimation for spectrum sharing. IEEE Trans. Commun. **64**(3), 973–986 (2016)
61. Raspberry Pi 4 Model B. [Online]. Available: https://www.raspberrypi.org/documentation/hardware/raspberrypi/bcm2711/rpi_DATA_2711_1p0_preliminary.pdf. Accessed 21 June 2019
62. H. Peng, X. Shen, Multi-agent reinforcement learning based resource management in MEC- and UAV-assisted vehicular networks. IEEE J. Sel. Areas Commun. **39**(1), 5894–5908 (2021)
63. Y. LeCun, L. Bottou, Y. Bengio, P. Haffner, Gradient-based learning applied to document recognition. Proc. IEEE **86**(11), 2278–2324 (1998)
64. W. Zhang, D. Yang, Y. Xu, X. Huang, J. Zhang, M. Gidlund, DeepHealth: a self-attention based method for instant intelligent predictive maintenance in industrial Internet of things. IEEE Trans. Ind. Inform. **17**(8), 5461–5473 (2021)

Chapter 4
Deterministic Transmission Scheduling for Time-Sensitive Federated Learning

4.1 Introduction

Through integrating emerging artificial intelligence techniques, sixth generation (6G)-enabled IIoT is expected to realize advanced industrial automation [1–3]. In such a scenario, an unprecedented amount of data is generated by dispersed IIoT devices, which needs to be distributively processed at the network edge due to delay requirements and privacy regulations [4, 5]. Over the past few years, a considerable amount of research efforts aiming at satisfying these requirements have led to the emergence of many novel distributed learning paradigms, which can leverage a large amount of dispersive data and computing power resources without transmitting raw data across the network [6–8]. In particular, federated learning has been proposed to enable collaborative DNN model training among devices and a centralized server via periodically exchanging model parameters (i.e., *FL traffic*), for the purpose of achieving high learning accuracy with preserved data privacy and reduced communication overhead [9, 10].

Operating FL over resource-constrained and time-sensitive IIoT faces a big challenge, that is, how to simultaneously transmit the FL traffic and traditional industrial traffic (e.g., time-triggered traffic, event-driven traffic, and best-effort traffic) over the same network and satisfy their diversified requirements [11]. In such a case, reliably aggregating the device-side model parameters in the centralized server is important to accelerate FL convergence [12]. time-sensitive networks, evolved from traditional industrial Ethernets, can provide multiple traffic flows with bounded end-to-end delay, jitter, and packet loss in local area networks, which is envisioned as a candidate for 6G-oriented ultra-reliable and low-latency communication techniques [13–15]. By introducing the synchronized time-scheduled approaches to traffic shaping mechanisms, such as cyclic queuing and forwarding proposed by IEEE 802.1Qch, TSN is endowed with flexible temporal resource scheduling capabilities to support deterministic traffic transmission [16].

© The Author(s), under exclusive license to Springer Nature Switzerland AG 2026

W. Zhang et al., *Intelligent Resource Scheduling in End-Edge-Cloud Networks*, Wireless Networks, https://doi.org/10.1007/978-3-032-07667-0_4

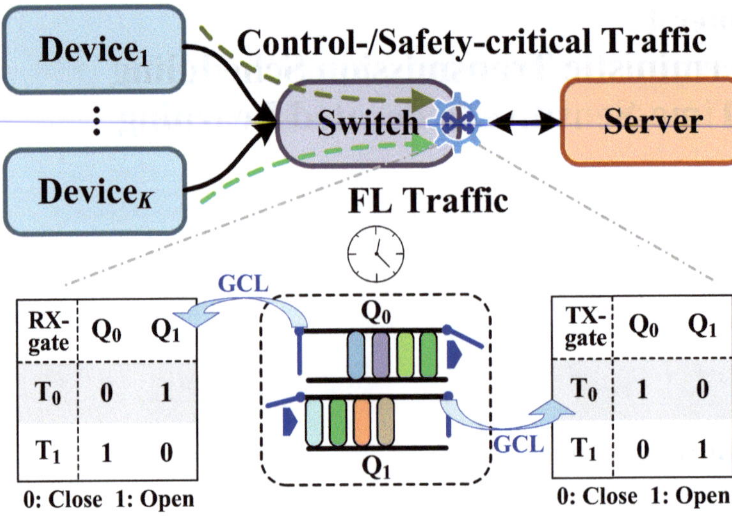

Fig. 4.1 CQF-based mechanism for deterministic traffic scheduling

As shown in Fig. 4.1, the CQF mechanism is able to provide each traffic flow with predictable and strictly bounded hop-by-hop transmission delay via periodically exchanging the receiving and sending statuses of Ping-Pong queues, which are deployed on each switch port operated by a couple of gates (i.e., RX-gate and TX-gate) [17]. The gate operations are controlled by a gate control list (GCL) to forward traffic, which takes the scheduling cycle as an operation cycle. Assume that the scheduling cycle is divided into two time slots, e.g., an even time slot T_0 and an odd time slot T_1. In T_0, the receiving queue Q_1 opens the RX-gate to receive traffic from upstream nodes, and the sending queue Q_0 opens the TX-gate to send traffic to downstream nodes. In T_1, the operations are performed in reverse. With the CQF mechanism, the TSN can support deterministic transmission for FL traffic between any two adjacent nodes, such that the FL convergence can be efficiently accelerated. However, when multiple IIoT devices participate in FL, all the devices interact with the centralized server in a parallel manner, resulting in inefficient resource preemption and increased transmission delay especially when the number of devices is large. To accelerate FL convergence, how can we schedule multiple traffic flows to facilitate FL over IIoT in an efficient way needs further investigation [18].

In this work, we propose a novel deterministic FL framework, named DetFed, which accelerates collaborative FL process via 6G-oriented ultra-reliable and low-latency communication techniques. The DetFed is composed of three layers, i.e., device, field, and factory layers. To provide the learning process with deterministic guarantee in delay, jitter, and packet loss, a wireless and wired integrated TSN is utilized to establish connections among the three layers, in which each switch is deployed with a CQF mechanism to support efficient model parameter distribution and aggregation. Aiming at optimizing the learning accuracy without affecting the transmission of burst traffic (e.g., safety-critical traffic), device-side updated models are periodically aggregated by their associated field servers or by a centralized factory server over the wireless and wired integrated TSN. In this way, the amount of transmitted model parameters among the field servers and the factory server can be effectively reduced, hence saving bandwidth occupancy on the industrial buses that can be used to support burst traffic transmission. As a result, more model parameters can be aggregated at the two-level server sides within the hierarchical DetFed, which significantly improves the learning accuracy.

To further improve the performance of DetFed, we propose a learning-based dynamic resource scheduling algorithm over the wireless and wired integrated TSN. The increase of the number of devices and network nodes may lead to intolerable time overhead in resource scheduling, because traditional optimization methods and heuristic schemes cannot adapt to network dynamics, which requires repeatedly resource scheduling calculations when a new device participates in the DetFed thus slowing down the FL convergence. To address this challenging issue, we investigate a resource scheduling problem in DetFed, which is formulated as a multi-objective stochastic optimization problem for maximizing the learning accuracy as well as the scheduling success ratio by optimizing temporal resource allocation for the wireless and wired integrated TSN. Since the objective function of the DetFed is implicit and the available time slots of the considered TSN in each FL round are temporally correlated, the problem is difficult to solve on-the-fly. We transform the problem into a Markov decision process formulation, and propose a dynamic resource scheduling algorithm, based on DRL, to solve the MDP. Extensive simulation results on real-world dataset with non-independent and identical distribution (non-IID) demonstrate that the proposed DetFed framework with the dynamic resource scheduling algorithm can significantly improve the learning accuracy as compared to the benchmarks. The main contributions of this chapter are summarized as follows:

- We present a hierarchical deterministic FL framework for the time-sensitive IIoT, in which FL can efficiently operate with bounded end-to-end delay, jitter, and packet loss guarantee via temporal resource scheduling for the wireless and wired integrated TSN;
- We formulate the multi-dimensional resource scheduling as a multi-objective stochastic optimization problem to maximize the FL performance over the TSN with deterministic service quality guarantee, which is then transformed into an MDP;
- We propose a D3QN-based resource scheduling algorithm to efficiently determine temporal resource allocation for FL traffic within the integrated TSN. After offline training, the proposed algorithm can facilitate FL convergence and adapt to network dynamics due to the traffic-by-traffic scheduling ability.

The reminder of this chapter is organized as follows. Related works are reviewed in Sect. 4.2. A deterministic FL framework is proposed in Sect. 4.3.1. Section 4.3 presents the system model, and formulates the optimization problem with multiple objectives. Section 4.4 transforms the formulated problem into an MDP, and proposes a dynamic resource scheduling algorithm. Section 4.5 provides the simulation results, followed by the concluding remarks in Sect. 4.6.

4.2 Related Work

4.2.1 FL in IIoT

FL has gained significant attention from both industry and academia due to the significant growth in data traffic generated at the network edge. It is regarded as a key enabling technology to achieve ubiquitous intelligence in envisioned 6G networks [19, 20]. Extensive research efforts have been devoted to resource management to accelerate FL convergence [21]. Chen et al. [22] propose a joint learning and communications framework for facilitating FL over wireless networks, in which spectrum resource allocation and user selection are jointly optimized to achieve a significant FL accuracy improvement. Wan et al. [23] aim to minimize the time and energy consumption during FL process through optimizing spectrum bandwidth allocation and the number of local updates per iteration. Dinh et al. [24] investigate a heterogeneous computing resource optimization problem, in which the local computation rounds and the global communication rounds are scheduled to speed up FL convergence at the cost of on-device computing power and spectrum resources. Taking into account the unreliable and resource-constrained wireless networks, Salehi et al. [25] proposed a novel FL framework to ensure extremely high transmission success probability via dynamically scheduling the participating devices for each FL round. Zhang et al. [26] study the joint optimization of the power allocation and learning rate adjustment to obtain the best estimate of the gradient updates while minimizing the impact of the communication error. In summary, most of these works adopt frequency division multiplexing as a main communication mode to support FL model parameter distribution and aggregation. Due to the limited spectrum resources, such a mode is generally difficult to guarantee deterministic delivery of FL traffic. Thus, in large-scale device scenarios, only a part of devices can be selected to participate in model training or only a few updated parameters can be uploaded to the global server, which significantly slows down FL convergence speed, degrades learning accuracy, and occupies network resources.

4.2.2 Deterministic Transmission Scheduling for FL

Recently, a line of research works are proposed to improve transmission determinacy [27–36]. Atallah et al. [37] studies an iterated integer linear programming based scheduling scheme to attain high scheduling scalability, in which time-triggered traffic is divided into multiple subsets and each subset is scheduled incrementally.

Yan et al. [38] proposes a CQF-based injection time planning algorithm to optimize the network throughput of time-sensitive traffic. Zhao et al. [39] considers a credit-based shaper to reduce the pessimism for the worst-case end-to-end delays of audio-video bridging traffic. More prominently, a few pioneering works focus on the co-transmission for multiple types of traffic. Yuan et al. [40] present a deadline monotonic scheduling algorithm to ameliorate the traffic transmission sequence of switch exports via dividing the queue priority into more levels, thereby satisfying the coordinated transmission requirements of time-triggered traffic and best-effort traffic. Huang et al. [17] propose a time-aware cyclic-queuing mechanism to support the co-transmission of cyclic flows and isochronous flows. To sum up, most of them are dedicated to supporting deterministic traffic transmission for traditional industrial applications (e.g., control- and/or safety-critical applications). How to provide industrial intelligent applications with deterministic transmission services, especially realizing the co-transmission of FL traffic and classical industrial traffic, is still a challenging issue needs to be solved.

Different from the existing works, this chapter focuses on facilitating efficient FL in time-sensitive IIoT which is to ensure the end-to-end deterministic transmission of FL traffic among devices and servers.

Furthermore, taking traffic time-varying property and network dynamics into account, we propose a learning-based temporal resource scheduling algorithm to accelerate FL convergence over a wireless and wired integrated TSN.

4.3 System Model of Time-Sensitive FL

4.3.1 Deterministic FL Framework

In the considered time-sensitive IIoT, we adopt a hierarchical FL framework enhanced by a two-level aggregation scheme to facilitate efficient model training [21]. As shown in Fig. 4.2, the framework consists of the device, field, and factory layers. The detailed descriptions of each layer in the framework are given as follows.

- **Factory Layer**: A global server and a central controller are deployed at the factory management center. The global server is responsible for initializing a set of model parameters at the beginning according to the application requirements and performing factory-level aggregation for updated model parameters (i.e., FL traffic) from distributed field servers every a few field aggregations. The controller is in charge of collecting network information and enforcing traffic scheduling decisions, as well as coordinating FL among three layers with deterministic guarantee of delay, jitter, and packet loss.
- **Field Layer**: Multiple TSN switches are deployed to provide deterministic transmission services for IIoT devices, in which a part of TSN switches endowed with a wireless access module are deployed to connect the devices into the factory network. The rest of the TSN switches are connected to the controller. In addition,

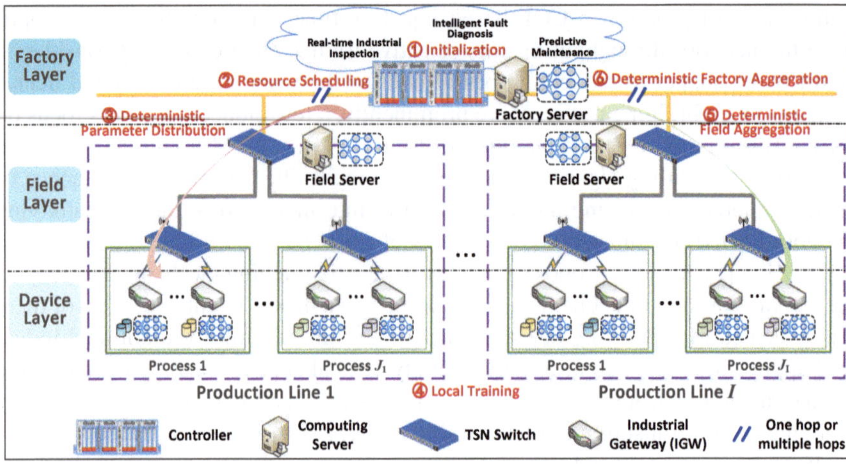

Fig. 4.2 A hierarchical deterministic FL framework for time-sensitive IIoT

each production line is equipped with a field server to aggregate model parameters from industrial IoT devices every FL rounds.

- **Device Layer**: IIoT devices endowed with certain storing and computing resources, such as industrial gateways, are distributively located in production lines within a smart factory. In each FL round, IGWs receive the global parameters that are transmitted from the factory server or field servers, and perform local training for further parameter updating.

The DetFed framework can support efficient FL with deterministic requirements through integrating delay-guaranteed TSN into a hierarchical aggregation scheme. On one hand, parameter aggregation and distribution are determinately coordinated among the three layers by accurate traffic scheduling. As such, more updated model parameters can be uploaded to the field servers or a factory server, thus accelerating FL convergence. On the other hand, model parameters are infrequently transmitted between the field and factory layers, thereby effectively avoiding bandwidth resource preemption with burst traffic in terms of factory bus.

Two-Level Deterministic Aggregation Scheme: In the framework, we consider a set of TSN switches to support deterministic data transmission. The set of wired TSN switches is denoted by $\mathcal{I} = \{1, 2, \ldots, I\}$. The set of wireless TSN switches is denoted by $\mathcal{J}_i = \{1, 2, \ldots, |\mathcal{J}_i|\}$, which is connected to a wired TSN switch $i \in \mathcal{I}$. Let $\mathcal{K}_j = \left\{1, 2, \ldots, \left|\mathcal{K}_j\right|\right\}$ be the set of IGWs associated with wireless TSN switch $j \in \mathcal{J}_i$. For IGW k, a set of data samples are stored locally, denoted by $\mathcal{S}_{i,j,k} = \{(\boldsymbol{u}_s, v_s)\}_s$. Here, $\boldsymbol{u}_s$ and v_s are the input and the label of each sample, respectively, s is the sample index. The training and testing sample sets of the IGWs are denoted by $\mathcal{S}_{i,j,k}^{\text{train}}$ and $\mathcal{S}_{i,j,k}^{\text{test}}$, respectively. The training and testing sample sets of the IGWs within the coverage of field server i are denoted by $\mathcal{S}_i^{\text{train}}$ and $\mathcal{S}_i^{\text{test}}$, respectively.

Algorithm 1: DetFed

◇ Initialization stage
Factory: Factory server initializes global DNN model parameter $\boldsymbol{\theta}^*$ with deterministic transmission requirements (i.e., end-to-end delay, jitter, and near-zero packet loss);
◇ Dynamic resource scheduling
Allocating temporal resources for $\mathcal{K}_j, \forall j \in \mathcal{J}_i, \forall i \in \mathcal{I}$ IIoT devices via dynamic resource scheduling to satisfy the deterministic transmission requirements;
for each $FL\ round\ r \in \mathcal{R}$ **do**
  ◇ Deterministic parameter distribution (**FL traffic**)
  Broadcast $\boldsymbol{\theta}_i$ from factory server or field servers to the connected IIoT devices via wireless and wired integrated scheduling;
  ◇ Local training (**heterogeneous computing power**)
  for each $TSN\ switch\ i \in \mathcal{I}$ *in parallel* **do**
    for each $device\ k \in \mathcal{K}_j, \forall j \in \mathcal{J}_i$ *in parallel* **do**
      for each *training sample* $s \in \mathcal{S}_{i,j,k}^{train}$ **do**
        Update the model parameters via a gradient descent method;
        $\boldsymbol{\theta}_{i,j,k} = \boldsymbol{\theta}_i - \eta \frac{\partial l(\boldsymbol{u}_s, v_s|\boldsymbol{\theta}_{i,j,k})}{\partial \boldsymbol{\theta}_{i,j,k}}$;
  ◇ Deterministic field aggregation (**wireless scheduling**)
  if $mod(r, r_g) \neq 0$ **then**
    Aggregate device-side model parameters by their associated field servers every FL round;
    $\boldsymbol{\theta}_i = 1/\left|\mathcal{S}_{i,j,k}^{test}\right| \sum_{i\in\mathcal{I}, j\in\mathcal{J}_i, k\in\mathcal{K}_j} \left|\mathcal{S}_{i,j,k}^{test}\right| \boldsymbol{\theta}_{i,j,k}$;
    $A(\boldsymbol{\theta}) = A_{fi}(\boldsymbol{\theta}_i)$;
  ◇ Deterministic factory aggregation (**wired scheduling**)
  if $mod(r, r_g) = 0$ **then**
    Aggregate field server-side model parameters by a factory server every a few FL rounds;
    $\boldsymbol{\theta} = 1/\left|\mathcal{S}^{test}\right| \sum_{i\in\mathcal{I}} \left|\mathcal{S}_i^{test}\right| \boldsymbol{\theta}_i$;
    $A(\boldsymbol{\theta}) = A_{fa}(\boldsymbol{\theta})$;
return $\boldsymbol{\theta}$, $A(\boldsymbol{\theta})$.

Here, we have $\mathcal{S}_{i,j,k}^{\text{train}} \cup \mathcal{S}_{i,j,k}^{\text{test}} = \mathcal{S}_{i,j,k}$, $\mathcal{S}_i^{\text{train}} \cup \mathcal{S}_i^{\text{test}} = \mathcal{S}_i$, and $\mathcal{S}_i = \cup_{j\in\mathcal{J}_i, k\in\mathcal{K}_j} \mathcal{S}_{i,j,k}$. A summary of important notations in this chapter is given in Table 4.1.

In the following, we elaborate the operation mechanism of the deterministic FL in detail, which is illustrated in Algorithm 1.

(1) ***Initialization Stage***: Once the factory management center receives an FL task, the factory server will instantiate a set of parameters $\boldsymbol{\theta}^*$ for obtaining the required learning model, such as a CNN-based industrial inspection model that can provide real-time quality detection services for industrial machinery.
(2) ***Multi-dimensional Resource Scheduling***: After completing the model initialization stage, the central controller then makes network resource scheduling decisions for FL round $r \in \mathcal{R}$ to support deterministic model parameter transmission. Here, the decisions include the amounts of time slots in wireless and wired TSN that are allocated to transmit model parameters among the IGWs, field servers, and a factory server. With these decisions, the FL can efficiently

Table 4.1 A summary of important notations

Notation	Description
$\mathcal{I}$	Set of wired TSN switches
$\mathcal{J}_i$	Set of wireless TSN switches connected to wired switch i
$\mathcal{K}_j$	Set of IGWs associated with wireless TSN switch j
$\mathcal{S}_{i,j,k}$	Set of data samples held by IGW k
$\mathcal{S}_i$	Set of data samples held by IGWs covered by field server i
$\mathcal{S}_{i,j,k}^{\text{train}}, \mathcal{S}_{i,j,k}^{\text{test}}$	Training and testing sample sets of the IGWs
$\mathcal{S}_i^{\text{train}}, \mathcal{S}_i^{\text{test}}$	Training and testing sample sets covered by field server i
$\boldsymbol{u}_s, v_s$	Input and label of each sample
$\boldsymbol{\theta}_{i,j,k}$	Locally trained parameters of each IGW
$\boldsymbol{\theta}_i$	Global model parameters of field servers or factory server
$l((\boldsymbol{u}_s, v_s))$	Training loss for sample $(\boldsymbol{u}_s, v_s)$ held by the IGWs
η	Learning rate for FL
r_g	Number of field aggregations
$A_i(\boldsymbol{\theta}_i)$	Accuracy on testing samples of the IGWs under field server i
$A_{fi}(\boldsymbol{\theta}_i)$	Accuracy of the updated model after field aggregations
$A_{fa}(\boldsymbol{\theta})$	Accuracy of the updated model after factory aggregations
t_k	FL traffic k
SC	Scheduling cycle
$\text{LCM}(\cdot)$	Least common multiple function
T, T_{num}	Time slot size and time slot number
$B^{\text{acc}}, B^{\text{core}}$	Wireless and wired link bandwidths
$\text{GCD}(\cdot)$	Greatest common divisor function
p_k^x	The xth packet of traffic t_k
$C_s, C_s^{\max}$	The occupied resources and maximum capacity for each slot
T_k^{e2e}, J_k^{e2e}	End-to-end delay and jitter for each traffic
PL_k^{e2e}	End-to-end packet loss for each traffic
H_k	The number of network nodes that traffic t_k traverses
$A(\boldsymbol{\theta}^r), S^r$	Learning accuracy and scheduling success ratio in round r
α	Temporal resource allocation decision
ξ	Weight coefficient for the objective function
s^t, a^t, z^t	Network state, decision, and agent reward
γ	Discount factor
$\pi_{\boldsymbol{\omega}}$	Policy parameterized by $\boldsymbol{\omega}$
$Q(\cdot)$	State-decision Q-value function
$V(\cdot), A(\cdot)$	Value function and advantage function
$L(\boldsymbol{\omega}), \nabla_{\boldsymbol{\omega}} L(\boldsymbol{\omega})$	Loss function and its gradient with regards to $\boldsymbol{\omega}$
y^d	Target Q-value

operate by orchestrating the dispersed IGWs, and achieve a higher accuracy through aggregating more model parameters with deterministic service quality guarantee.

(3) ***Deterministic Parameter Distribution***: Given the resource scheduling decisions, the aggregated model parameters can be distributed from the field servers or a factory server to the IGWs in each FL round. Due to adopting the deterministic traffic scheduling mechanism, the parameter distribution can have a high schedulability, such that more IGWs can receive the distributed model parameters to participate in the current FL round.

(4) ***Local Training***: When the distributed model parameters $\boldsymbol{\theta}_i, \forall i \in \mathcal{I}$ are received, the IGWs can start local training using their sample sets for one round. To speed up convergence rate, we adopt a stochastic gradient descend method to iterate the model parameters, that is, $\boldsymbol{\theta}_{i,j,k} = \boldsymbol{\theta}_i - \eta \partial l((\boldsymbol{u}_s, v_s) \mid \boldsymbol{\theta}_{i,j,k})/\partial \boldsymbol{\theta}_{i,j,k}$. Here, $\boldsymbol{\theta}_{i,j,k}$ is the locally trained parameters of each IGW, η is the learning rate, and $l((\boldsymbol{u}_s, v_s) \mid \boldsymbol{\theta}_{i,j,k})$ is the training loss with regards to each sample held by the IGWs.

(5) ***Deterministic Field Aggregation***: Once completing the local training, field servers will asynchronously aggregate the trained model parameters from the IGWs via deterministic resource scheduling, namely, $\text{mod}(r, r_g) \neq 0$. Meanwhile, the global model parameters $\boldsymbol{\theta}_i$ can be updated through a FedAvg algorithm that performs a weighted average calculation for all the aggregated parameters, i.e., $\boldsymbol{\theta}_i = 1/\left|\mathcal{S}_i^{\text{test}}\right| \sum_{i \in \mathcal{I}, j \in \mathcal{J}_i, k \in \mathcal{K}_j} \left|\mathcal{S}_{i,j,k}^{\text{test}}\right| \boldsymbol{\theta}_{i,j,k}$. Here, $\left|\mathcal{S}_{i,j,k}^{\text{test}}\right|$ and $\left|\mathcal{S}_i^{\text{test}}\right|$ are the numbers of testing samples locally stored at each IGW and at all IGWs associated with field server i. Thus, the learning accuracy of the updated model is given by

$$A_{fi}(\boldsymbol{\theta}_i) = \frac{1}{I} \sum_{i \in \mathcal{I}} A_i(\boldsymbol{\theta}_i), \tag{4.1}$$

where $A_i(\boldsymbol{\theta}_i)$ denotes the learning accuracy on testing samples of the IGWs that are associated with field server i in terms of its aggregated parameters $\boldsymbol{\theta}_i$, and $A_i(\boldsymbol{\theta}_i) = 1/\left|\mathcal{S}_i^{\text{test}}\right| \sum_{s \in \mathcal{S}_i^{\text{test}}} a((\boldsymbol{u}_s, v_s) \mid \boldsymbol{\theta}_i)$. Here, $a((\boldsymbol{u}_s, v_s) \mid \boldsymbol{\theta}_i)$ is the learning accuracy on each testing sample held by the IGWs. Then, the aggregated model parameters of field servers will be distributed to their associated IGWs for a new round update.

(6) ***Deterministic Factory Aggregation***: After r_g field aggregations, the factory server will aggregate model parameters from field servers via deterministic resource scheduling for the considered TSN networks, namely, $\text{mod}(r, r_g) = 0$. Then, the global model parameters can be updated via the FedAvg algorithm, i.e., $\boldsymbol{\theta} = 1/\left|\mathcal{S}^{\text{test}}\right| \sum_{i \in \mathcal{I}} \left|\mathcal{S}_i^{\text{test}}\right| \boldsymbol{\theta}_i$. The learning accuracy can be described as

$$A_{fa}(\boldsymbol{\theta}) = \frac{1}{|\mathcal{S}^{\text{test}}|} \sum_{s \in \mathcal{S}^{\text{test}}} a((\boldsymbol{u}_s, v_s) \mid \boldsymbol{\theta}), \tag{4.2}$$

where $\mathcal{S}^{\text{test}} = \cup_{i \in \mathcal{I}} \mathcal{S}_i^{\text{test}}$, $|\mathcal{S}^{\text{test}}|$ is the total number of testing samples stored at all the IGWs, and $a((\boldsymbol{u}_s, v_s) \mid \boldsymbol{\theta})$ is the learning accuracy on each testing sample of all the IGWs with parameters $\boldsymbol{\theta}$ aggregated by the factory server.

In the considered scenario, the proposed DetFed framework adopts *synchronous FL paradigm*, which is operated round-by-round. In each FL round, due to the heterogeneous computing resources of IoT devices, the delay for local training is different for each device. Whenever the local training stage is completed, each IoT device will send a signaling to its associated field server or the factory server. Once the signaling from the last participating IoT device is received, the field server will synchronously aggregate model parameters from their participating devices every FL round, and the factory server will aggregate model parameters from the edge servers every a few FL rounds. To maintain time synchronization within the considered network scenario, the clock synchronizing mechanism *IEEE 1588* is deployed at the TSN switches and IoT devices. As such, the FL traffic of heterogeneous IoT devices can be scheduled synchronously.

Computation Complexity Analysis: Different from the conventional FL framework that aggregates model parameters only by one centralized server, a set of additional field servers are deployed in the proposed DetFed framework to aggregate the uploaded model parameters from their connected IoT devices. The computation complexity of both frameworks is mainly composed of local training and model aggregation. For the local training stage, both frameworks have the same computation complexity when they train the same DNN model. For the model aggregation stage, the computation complexity of the DetFed is slightly higher than the conventional FL. However, the model aggregation can be implemented via a low-complexity algorithm, e.g., FedAvg, in which model parameters are simply aggregated via a weighted average method. We assume that the training computation complexity is denoted as $\mathcal{O}^{\text{tra}}(\cdot)$, and that the computation complexity of operating the FedAvg is denoted as $\mathcal{O}^{\text{agg}}(\cdot)$. The aggregation computation complexity of the DetFed is denoted as $(I+1) \times \mathcal{O}^{\text{agg}}(\cdot)$. Here, I is the number of field servers, and 1 indicates to the factory server. Thus, the total computation complexity of the DetFed is $\mathcal{O}^{\text{tra}}(\cdot) + (I+1) \times \mathcal{O}^{\text{agg}}(\cdot)$. Here, we have $\mathcal{O}^{\text{tra}}(\cdot) \gg (I+1) \times \mathcal{O}^{\text{agg}}(\cdot)$.

4.3.2 FL Traffic Model

As previously described, the FL is a long-term and iterative process, in which model parameters are interacted alternately among IGWs, field servers, and a factory server. In such scenarios, whether the model parameters can reliably transmit among the three layers is especially important to accelerate FL convergence. For parameter distribution, if the global model parameters can be distributed to more IGWs, the FL can take full advantage of geographically dispersed data to train the global model. For parameter aggregation, if more local trained parameters can be uploaded to the field servers or the factory server, the global model can achieve a faster convergence. Thus, to ensure the deterministic parameter transmission during the long-term learning process, we operate the hierarchical FL over a TSN-enhanced network scenario, and define the model parameters as *FL traffic* with deterministic end-to-end transmission requirements need to be guaranteed.

In the considered scenario, the transmitted FL traffic requires the bounded end-to-end delay, jitter, and packet loss. As such, each FL traffic can be described as

$$\begin{aligned} &t_k \triangleq \{\text{id}, \text{cycle}, \text{src}, \text{dst}, \text{path}, d_{e2e}, j_{e2e}, p_{\text{num}}, p_{\text{size}}, \text{st}\}, \\ &\forall t_k \in K, k = 1, 2, \ldots, |K|, \end{aligned} \tag{4.3}$$

where each traffic t_k has ten features, including identifier number, traffic cycle, source and destination addresses, preset transmission path, end-to-end delay and jitter requirements, packet number and size, and sending time that is a decision variable needs to be made by a resource scheduling algorithm. Here, the sending time of a traffic means the timestamp when the traffic is sent from a device's network interface card (NIC). In addition, $|K| = \sum_{i \in \mathcal{I}} \sum_{j \in \mathcal{J}_i} |\mathcal{K}_j|$ denotes the total number of FL traffic in each FL round.

4.3.3 Wireless and Wired Integrated Scheduling Model

As stated earlier, the FL operates round-by-round within the hierarchical framework. To ensure a high convergence efficiency, the FL traffic should be transmitted among the three layers. Taking into account the mobility, the IIoT devices are connected to the corresponding wireless TSN switches endowed with a wireless access module [41]. In addition to that, FL traffic are forwarded by a set of wired TSN switches. In each TSN switch, we deploy the CQF-based shaper to satisfy the deterministic transmission requirements, which is endowed with a queuing mechanism for shaping FL traffic to provide each traffic with bounded delay, jitter, and packet loss [42]. To implement high-performance TSN, network adapters that are programmable and support TSN functionality is required at each IoT device of the communication link. In this work, the industrial IoT devices are directly connected to the TSN switches and each IoT device is equipped with a network adapter (e.g., Intel I225-V) to support FL traffic transmission. In the following, several important constraints are defined:

Traffic Scheduling Cycle: In the CQF model, a pair of Ping-Pong queues are introduced at each switch port, and the open (i.e., sending) and closed (i.e., receiving) statuses of Ping-Pong queues are periodically exchanged to forward traffic that is controlled by GCLs, such that the hop-by-hop bounded delay can be guaranteed to provide FL traffic with deterministic transmission services [37]. Since the GCL calculations need to be cyclically executed for all FL traffic, scheduling cycle SC equals to the least common multiple of the set of traffic cycles, namely,

$$\text{SC} = \text{LCM}(K[\text{cycles}]), \tag{4.4}$$

where SC should be a constant value, LCM$(\cdot)$ is the function to return the least common multiple value, and K[cycles] denotes the set of cycles of all FL traffic.

Time Slot Size: As defined by CQF, a scheduling cycle SC should be divided into several time slots with the same duration T. Thus, the number of time slots in each scheduling cycle is $T_{\text{num}} = \text{SC}/T$. To avoid data packets conflicting with each other in the same time slot, the value of T should be within $[T^{\min}, T^{\max}]$.

For the minimum value $T^{\min}$, it means the considered integrated networks should be able to accommodate the largest size of the FL traffic, i.e.,

$$T^{\min} = \tau_{\text{send}} + \tau_{\text{prop}} + \tau_{\text{proc}} + \tau_{\text{queue}} + \tau_{\text{sync}}, \tag{4.5}$$

where

$$\tau_{\text{send}} = \begin{cases} \frac{\max(K[\text{sizes}])}{B^{\text{acc}}}, & \text{if ACC} = \text{True}, \\ \frac{\max(K[\text{sizes}])}{B^{\text{core}}}, & \text{otherwise}, \end{cases} \tag{4.6}$$

where τ_{send}, τ_{prop}, τ_{proc}, τ_{queue}, and τ_{sync} denote the packet sending, propagating, processing, queuing, and synchronizing delays between any two adjacently connected nodes, respectively. In addition, $\max(\cdot)$ is the function to return the largest value, K[sizes] denotes the set of data sizes of all FL traffic, and $B^{\text{acc}} = R^{\text{acc}} \log_2\left(1 + p_{i,j}|h_{i,j,k}|^2/\mu^2\right)$, $\forall i \in \mathcal{I}, j \in \mathcal{J}_i, k \in \mathcal{K}_j$ and B^{core} denote the wireless and wired link bandwidths, respectively. Here, ACC is True indicates that the traffic is transmitted between an IGW and a wireless TSN switch. Also, R^{acc} is the frequency band of the wireless networks, $p_{i,j}$ is the transmission power, $\left|h_{i,j,k}\right|^2$ is the channel gain between wireless TSN switch j and IGW k, and μ^2 is the Gaussian noise.

For the maximum value $T_{\max}$, it means the size of time slots should be larger than the greatest common divisor of the set of traffic cycles, i.e.,

$$T^{\max} = \text{GCD}(K[\text{cycles}]), \tag{4.7}$$

where GCD$(\cdot)$ is the function to return the greatest common divisor value. In addition, the scheduling cycle should be divisible by T, i.e.,

$$\text{SC} \;\%\; T = 0. \tag{4.8}$$

Sending Time of FL Traffic: In each FL round, FL traffic is transmitted among IGWs, field servers, and a factory server bottom up or top down. In either case, however, the sending time of each FL traffic should be smaller than its traffic cycle. In this way, packets of different traffic cycles can be prevented from preempting limited NIC resources in the meantime. Hence,

$$0 \leq t_k[\text{st}] < t_k[\text{cycle}], \forall k \in K. \tag{4.9}$$

Note that the sending time $t_k[\text{st}]$, namely injecting packets of traffic k into which time slots within each scheduling cycle, is determined by a centralized network controller that is in charge of global resource scheduling. In addition, similar to the scheduling cycle, the traffic cycle should also be divisible by T, i.e.,

$$t_k[\text{cycle}] \ \% \ T = 0, \forall k \in K. \tag{4.10}$$

FL Traffic Transmission: With the sending time decisions, the occupied temporal resources for each time slot of the current scheduling cycle can be calculated, that is,

$$C_s = \sum_{k \in K} \sum_{x \in t_k[p_{\text{num}}]} \Omega_s(p_k^x) \cdot t_k[p_{\text{size}}], \forall s \in [1, T_{\text{num}}], \tag{4.11}$$

where $\Omega_s(\cdot)$ is an indicator function, and p_k^x is the xth packet of traffic t_k. If $\Omega_s(p_k^x) = 1$, the packet p_k^x is injected into time slot s, otherwise 0. In particular, each packet can only be injected into no more than one time slot, i.e.,

$$\sum_{s \in T_{\text{num}}} \Omega_s(p_k^x) \leq 1, \forall k \in K, x \in [1, t_k[p_{\text{num}}]]. \tag{4.12}$$

Note that since all packets stored in the sending buffer need to be sent before the start of the next time slot, the size of the total packets within each time slot cannot exceed its maximum capacity. Hence,

$$0 \leq C_s \leq C_s^{\max}, \forall s \in [1, T_{\text{num}}], \tag{4.13}$$

where $C_s^{\max}$ is the maximum capacity of time slot s. Note that the deterministic transmission guarantee in TSN is to reserve bandwidth resources for traffic in advance. For FL traffic, the traffic information can be available in advance and a temporal resource allocation policy is determined accordingly for transmission. However, for burst traffic, it is often difficult to determine a dynamic resource allocation policy to support real-time traffic due to information uncertainty. To ensure that the reserved resources are sufficient while avoiding resource wasting, we consider to reserve bandwidth in a statistical way by considering the long-term traffic features to accommodate the burst traffic, and the rest of link bandwidth resources are utilized by FL traffic. As such, FL traffic can be co-transmitted with burst traffic within the considered TSN scenario.

Multiple Traffic Requirements: Based on the above definitions, the end-to-end delay T_k^{e2e}, jitter J_k^{e2e}, and packet loss PL_k^{e2e} for each traffic can be obtained, namely,

$$T_k^{e2e} = (H_k + 1) \cdot T, \forall k \in K,$$
$$J_k^{e2e} = \left| p_k^x[d_{e2e}] - \frac{1}{t_k[p_{\text{num}}]} \sum_{x \in t_k[p_{\text{num}}]} p_k^x[d_{e2e}] \right|, \forall k \in K, \tag{4.14}$$
$$\text{PL}_k^{e2e} = \left| t_k[p_{\text{num}}] - \sum_{x \in t_k[p_{\text{num}}]} \sum_{s \in T_{\text{num}}} \Omega_s(p_k^x) \right|, \forall k \in K,$$

where H_k is the number of network nodes that traffic t_k traverses. Note that the jitter of each traffic is the delay variance of all its packets. Here, $p_k^x[d_{e2e}]$ denotes the end-to-end delay of the xth packet of traffic t_k, and $1/t_k[p_{\text{num}}] \sum_{x \in t_k[p_{\text{num}}]} p_k^x[d_{e2e}]$ is the average delay of all packets for traffic t_k. The jitter of each traffic requires the gap between both values do not exceed the required jitter. The definition of packet loss is the same. Only when the above requirements are satisfied simultaneously, the FL traffic then can be transmitted with a deterministic service quality guarantee.

4.3.4 *Multi-objective Optimization Problem Formulation*

Since the learning accuracy of server-side models and the scheduling capability of the integrated networks are mutually reinforcing, both metrics should be jointly considered to maximize the overall system performance. In particular, the presented FL framework requires making the best time slot allocation decisions for deterministic model parameter transmission over the wireless and wired integrated TSN. To this end, we formulate the multi-objective resource scheduling problem as:

$$\mathcal{P} : \max_{\boldsymbol{\alpha}} \ \xi \left| A(\boldsymbol{\theta}^r) \right| + (1 - \xi) \left| S^r \right|$$

$$\text{s.t.} \quad T_k^{e2e} \le t_k[d_{e2e}], \forall k \in K, \tag{4.15a}$$
$$J_k^{e2e} \le t_k[j_{e2e}], \forall k \in K, \tag{4.15b}$$
$$\text{PL}_k^{e2e} \le \delta, \forall k \in K, \tag{4.15c}$$
$$T \in [T^{\min}, T^{\max}], \tag{4.15d}$$
$$(4.4), (4.8), (4.9), (4.10), (4.12), \text{ and } (4.13),$$

where $\boldsymbol{\alpha} = \{\alpha_{i,j,k}\}_{\forall i \in \mathcal{I}, j \in \mathcal{J}_i, k \in \mathcal{K}_j}$ denotes the temporal resource allocation decision, and $\alpha_{i,j,k} \in [1, T_{\text{num}}]$ is a discrete value which corresponds to the sending time of each FL traffic defined in Eq. (4.9). In addition, $\left| A(\boldsymbol{\theta}^r) \right| \in [0, 1], \forall r \in \mathcal{R}$ denotes the learning accuracy of the global model with respect to the aggregated parameters $\boldsymbol{\theta}^r$, and $|S^r| \in [0, 1]$ denotes the scheduling success ratio in FL round r which is the ratio of the number of successfully scheduled traffic to the number of all traffic. In the proposed DetFed framework, scheduling success ratio and learning accuracy are both important to accelerate FL convergence. A high scheduling success ratio is helpful to aggregate more updated model parameters from the distributed IoT devices, and a higher learning accuracy indicates a better collaborative learning performance. This section aims to jointly optimize both objective components to efficiently obtain the optimal FL model parameter $\boldsymbol{\theta}^{\star}$. Thus, the weight coefficient $\xi \in (0, 1)$ is to balance the importance between both objective components. Constraints (4.15a), (4.15b), and (4.15c) guarantee the traffic requirements can be satisfied to provide FL traffic with deterministic transmission support. Here, δ is a tiny constant value. Constraint (4.15d) ensures the time slot duration T of the considered integrated network is set within a valid range, and constraints (4.4), (4.8), (4.9), (4.10), (4.12), and (4.13) guarantee proper network resource scheduling.

The *relationship* between the resource allocation decision and the objective function is as follows. An appropriate decision α is helpful to distribute the global FL model from the field servers or factory server to more IoT devices and to aggregate more updated FL models from the IoT devices, thereby obtaining a higher scheduling success ratio $|S^r|$. Moreover, with decision α, more dispersed data held by the IoT devices can be utilized to sufficiently optimize the FL model parameters $\boldsymbol{\theta}$ in the local model training stage, which further improves learning accuracy $\left|A(\boldsymbol{\theta}^r)\right|$.

Problem $\mathcal{P}$ is a *multi-objective stochastic* optimization problem. The problem is "multi-objective" because the learning accuracy and scheduling success ratio are jointly considered to achieve a faster FL convergence. The problem is "stochastic" due to the resource scheduling decisions are determined in presence of temporal dynamics of available time slot resource during the FL process. In addition, the first term of $\mathcal{P}$ (i.e., learning accuracy) is expressed by an implicit function, and it is difficult to accurately characterize the mapping relationship between the decision variables and objective function, which further complicates decision making process. In particular, the available time slot capacity constraint is coupled with temporal resource scheduling decisions in terms of time and the amount of the scheduled traffic, which however have to be determined without foreseeing future network dynamics. Furthermore, according to the definitions of optimization components, the objective function is non-convex. Since optimizing variable α is integer, problem $\mathcal{P}$ is an integer optimization problem with a non-convex objective function. As such, the optimal variable α cannot be directly obtained via existing optimization methods. To approach the optimal solution, we consider DRL-based approaches to design a dynamic resource scheduling algorithm, which can make the optimal temporal resource allocation decisions and thus solving the formulated problem in an efficient way.

4.4 Dynamic Resource Scheduling Scheme for FL Traffic

In this section, we first transform the multi-objective optimization problem into an MDP, and then propose a model-free DRL-based resource scheduling algorithm, named DynamicSched, to solve it. The computational complexity analysis of the DynamicSched is also provided.

4.4.1 MDP Transformation

MDP is a temporal decision making process, which is generally utilized to simulate the stochastic policies and rewards of learning agents in the environment with Markov property. Here, Markov property refers to that given the present state and all the past states of a random process, the conditional probability distribution of the next state depends only on the present state. In this section, the formulated problem $\mathcal{P}$ is to design a dynamic resource scheduling policy to maximize the learning accuracy of server-side models and the scheduling capability of the considered networks while satisfying constraints, which falls into the MDP. To optimize FL performance over the time-sensitive industrial IoT, we reformulate the resource scheduling problem $\mathcal{P}$ as an MDP. As shown in Fig. 4.3, we adopt a fully-centralized architecture proposed by IEEE 802.1Qcc in the control plane, and the central controller is modeled as an agent [43]. In each FL round, the agent can observe the traffic information and network state $s^t, t \in [1, |K|]$, and take temporal resource allocation decision $a^t, t \in [1, |K|]$ traffic-by-traffic. Here, $|K| = \sum_{i \in \mathcal{I}} \sum_{j \in \mathcal{J}_i} \mathcal{K}_j$ is the total number of FL traffic in each FL round. This means that in the MDP, the time index t denotes a traffic index from the traffic set that needs to be scheduled. To learn an optimal policy $\boldsymbol{\pi}^*_{\omega}$ parameterized by $\boldsymbol{\omega}$, that can accurately map state $s^t, t \in [1, |K|]$ to decision $a^t, t \in [1, |K|]$, the corresponding reward $z^t, t \in [1, |K|]$ is fed back from the network environment, and the state is transformed into new state $s^{t+1}, t \in [1, |K|]$ accordingly [44]. In the MDP, three core elements, i.e., state, action (i.e., decision), and reward, are defined as follows.

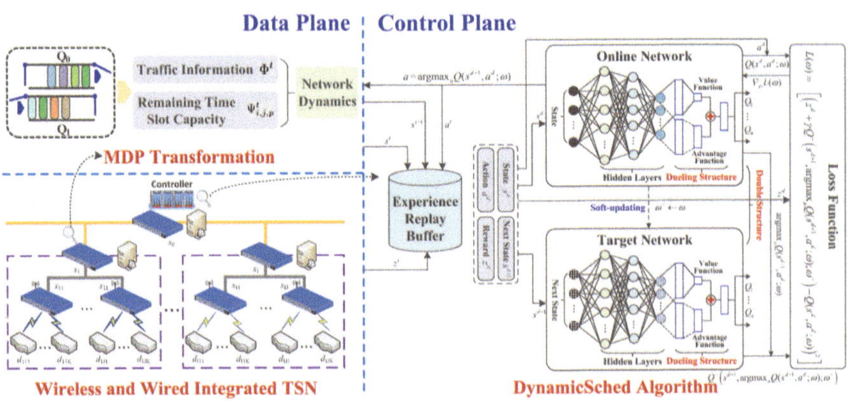

Fig. 4.3 An illustration of the proposed DynamicSched algorithm

- **State**: In FL round r, the agent collects traffic information Φ^t and the remaining time slot capacities $\Psi^r_{i,j,p}$ from the IIoT devices and TSN switches. Then, the state $s^t \in \mathcal{S}$ of the considered TSN can be defined as

$$s^t = \{\Phi^t, \Psi^t_{i,j,p}\}, \forall k \in \mathcal{K}_j, j \in \mathcal{J}_i, i \in \mathcal{I}, \tag{4.16}$$

where $\Phi^t \triangleq \{t_k\}_{k \in \mathcal{K}_j}$ denotes the features of traffic t_k that is received at round r, and $\Psi^t_{i,j,p}$ denotes the remaining capacities of each receiving queue on port $p \in P$ for each time slot within a scheduling cycle.
- **Action**: Based on the observed state and the parameterized policy, the agent can make the temporal resource allocation decisions to determine which time slot is scheduled to transmit t_k, thereby supporting deterministic transmission for FL traffic. Namely,

$$a^t = \begin{cases} \arg\max_{a^t_\varepsilon \in \mathcal{A}} Q(s^t, a^t_\varepsilon), & \text{if } \varepsilon \geq \varepsilon_0 - N, \\ a^t_{\text{rand}}, & \text{otherwise}, \end{cases} \tag{4.17}$$

Each action element is an integer value reshaped into a range of $[0, t_k[\text{cycle}]/\text{SC} - 1]$. The action indicates which time slot the FL traffic should be injected. Note that each traffic can only be injected into one time slot to be transmitted. Here, $\varepsilon_0 \in (0, 1]$ is an initial probability value, and N is a decay factor for each learning step.
- **Reward**: Once the agent takes action a^t, it will obtain a reward to evaluate how good the action is under state s^t. Aiming at maximizing the learning accuracy and scheduling success ratio simultaneously, the reward function of the DynamicSched algorithm can be given by

$$z^t = \begin{cases} V_{\text{succ}} + Z_{fi}(\boldsymbol{\theta}_i) \cdot \mathbb{1}\{\text{mod}(r, r_g) \neq 0\}, & \text{if success}, \\ V_{\text{succ}} + Z_{fa}(\boldsymbol{\theta}) \cdot \mathbb{1}\{\text{mod}(r, r_g) = 0\}, & \text{if success}, \\ V_{\text{fail}}, & \text{if failed}, \end{cases} \tag{4.18}$$

where

$$Z_{fi}(\boldsymbol{\theta}_i) = \xi \left|A_{fi}(\boldsymbol{\theta}^r_i)\right| + (1-\xi)\left|S^r\right|, \forall i \in \mathcal{I}, \tag{4.19}$$

and

$$Z_{fa}(\boldsymbol{\theta}) = \xi \left|A_{fa}(\boldsymbol{\theta}^r)\right| + (1-\xi)\left|S^r\right|. \tag{4.20}$$

There are three cases for the reward. When the current FL traffic is scheduled successfully, the agent will obtain a positive reward, i.e., $V_{\text{succ}} + Z_{fi}(\boldsymbol{\theta}_i), \forall i \in \mathcal{I}$ or $V_{\text{succ}} + Z_{fa}(\boldsymbol{\theta})$, which are consistent with the two-level aggregation scheme in DetFed. Here, V_{succ} is a positive constant value, and $A_{fi}(\boldsymbol{\theta}^r_i), \forall i \in \mathcal{I}$ and $A_{fa}(\boldsymbol{\theta}^r)$ denote the learning accuracies on the model parameters after field aggregation (i.e., $\mathbb{1}\{\text{mod}(r, r_g) \neq 0\}$) and factory aggregation (i.e., $\mathbb{1}\{\text{mod}(r, r_g) = 0\}$), which are given by Eqs. (4.1) and (4.2), respectively. On the contrary, the agent will obtain a negative constant value V_{fail} to penalize the inappropriate decisions. Obviously, the reward is related to the average learning accuracy and scheduling success ratio in the current FL round. With the obtained reward, the parameterized policy π_ω can be optimized during the training stage until algorithm converges.

In the MDP, the goal of the agent is to find the optimal temporal resource allocation policy $\boldsymbol{\pi}_\omega^\star \in \Pi$ that can maximize cumulative discounted reward, namely,

$$\mathcal{P}' : \max_{\pi_\omega \in \Pi} \mathbb{E}\left[\sum_{\rho=0}^{\infty} \gamma^\rho z^{t+\rho}|(s^t, a^t; \boldsymbol{\pi}_\omega)\right]$$

$$\text{s.t.} \quad (4.4), (4.8), (4.9), (4.10), (4.12), (4.13), \text{ and } (4.15a) - (4.15c), \qquad (4.21\text{a})$$

where $\gamma \in [0, 1]$ denotes the discount factor. Note that policy $\boldsymbol{\pi}_\omega$ specifies the way the central controller adopted to schedule temporal resources according to the observed states in each scheduling cycle.

4.4.2 D3QN-Based Resource Scheduling Algorithm

The motivation to solve the MDP problem via deep RL algorithms are due to the following two aspects: (1) In the considered TSN scenario, it is difficult to obtain the complete information on state transition probabilities which is required to apply MDP algorithms (e.g., Relative Value Iteration) due to large state and action spaces; (2) The problem size is large, the convergence of conventional value iteration algorithms is time consuming [45]. However, deep RL algorithms can solve the MDP problem by offline training and online implementation with a low latency even when the problem size is large, which can be more efficiently applied in practical systems. Compared to traditional DQN-based algorithms, the D3QN algorithm can significantly reduce overestimation for resource scheduling decisions. Moreover, the D3QN algorithm can identify the correct resource scheduling decision more quickly during policy evaluation and achieve much better decision performance via a *dueling* DQN module. The above designs contribute to better approximation of the optimal state values via frequent updating of the value stream, thus improving the stability of resource scheduling optimization especially when the number of scheduling decisions is large. Thus, we consider the D3QN algorithm to solve the MDP. As shown in Fig. 4.3, we propose a learning-based dynamic resource scheduling algorithm based on D3QN, named DynamicSched. In the algorithm, a *double* DQN and a *dueling* DQN modules are adopted to overcome the decision overestimation problem and achieve much better performance [46, 47]. For the double DQN module, an online network is to evaluate the $\varepsilon - greedy$ policy, and a target network is to estimate its value. With this design, the original *max* operation in the target Q-value estimation of DQN algorithms can be decomposed into decision selection and decision evaluation, therefore effectively reducing overestimation for decisions. For the dueling DQN module, the lower layers of online and target networks adopt fully-connected structures parameterized by φ and φ', respectively, while the following layers of both networks are endowed with two streams of FC layers instead of the original single stream. Here, the two streams represent the value function $V(s^t; \omega, \varphi')$ and advantage function $A(s^t, a^t; \omega, \varphi)$ that sharing a common feature learning layers, that is, the lower layers of online and target networks. In the output aggregating layer, the two streams are combined to produce a single state-decision Q-value function, namely,

$$Q(s^t, a^t; \omega, \varphi, \varphi') = V(s^t; \omega, \varphi') + \left(A(s^t, a^t; \omega, \varphi) - \frac{1}{|\mathcal{A}|} \sum_{a^t} A(s^t, a^t; \omega, \varphi) \right), \tag{4.22}$$

where $V(\cdot)$ is an one-dimension scalar, and $A(\cdot)$ is an $|\mathcal{A}|$-dimension vector. Here, $|\mathcal{A}|$ is the dimension of decision a^t. With this improvement, the value and advantage functions can be estimated separately, such that the effect of state s^t and decision a^t on Q-value estimation can be taken into account simultaneously and the correct decision can be identified more quickly during policy evaluation. Note that the better decision a^t, the greater the advantage. For notation simplicity, we omit φ and φ' in the following context.

In this subsection, we illustrate the D3QN-based dynamic resource scheduling algorithm as the following steps, which is presented in Algorithm 2.

(1) *Algorithm Initialization*: Once the training stage is started, the parameters of the online and target networks of the centralized learning agent will be initialized and then updated step-by-step. Note that only one FL traffic is scheduled in each learning step, and thus the size of the learning step equals to the total number of the traffic that needs to be scheduled. Additionally, an experience replay buffer $\mathcal{B}$ is instantiated with capacity D to store the sampled experiences that can be utilized to update the algorithm parameters. Then, the considered network environment is reset to generate an initial state s^1.
(2) *Experience Sampling*: When the initialization stage is completed, the learning process of the agent can start by sampling interacted experiences from the network environment and storing them into replay buffer $\mathcal{B}$, which can be denoted as a five-element tuple, i.e., $\{s^t, a^t, z^t, s^{t+1}, \text{done}\}$. Specifically, the learning agent makes temporal resource allocation decision a^t with regard to current state s^t. Note that decision $a^t = \alpha_{i,j,k}, \forall i \in \mathcal{I}, j \in \mathcal{J}_i, k \in \mathcal{K}_j$ is made by alternating two policies, i.e., $\varepsilon - greedy$ policy and *random* policy. Then, decision a^t is executed to allocate time slot resources for current FL traffic t_k, and an immediate reward z^t can be obtained from the network environment via performing Algorithm 1 in Chap. 3. As such, the network state will be transformed into new state s^{t+1}, and

Algorithm 2: DynamicSched

Input: Traffic information and remaining time slot capacities
$\{\Phi^t, \Psi^t_{i,j,p}\}, \forall i \in \mathcal{I}, j \in \mathcal{J}_i, p \in \mathcal{P}$;
Output: Optimal temporal resource allocation policy $\pi^\star_\omega$ that can make decision a^t traffic-by-traffic;
Initialize online and target networks with parameters ω and ω^-, and initialize experience replay buffer $\mathcal{B}$ to capacity D;
for *episode* $e \in [1, |E_{num}|]$ **do**
  Reset network environment and obtain initial state s^1;
  for *step* $t \in [1, |K|]$ **do**
    ◇ Environment sampling
    Obtain random decision a^t_{rand} with probability ε, otherwise obtain $\text{argmax}_a Q^\star(\phi(s^t), a^t_\varepsilon; \pi_\omega)$ with $\epsilon - greedy$ policy;
    Execute decision a^t (i.e., allocate time slot resource for current FL traffic t_k) and obtain an immediate reward z^t;
    Transform to new state s^{t+1} to obtain the available time slot capacities of the considered wireless and wired integrated TSN networks for the next traffic;
    Store $\{s^t, a^t, z^t, s^{t+1}, done\}$ into $\mathcal{B}$, and replace the oldest experiences if $|\mathcal{B}| > D$;
    ◇ Neural network training
    Sample a random minibatch of D_b experiences $\{s^d, a^d, z^d, s^{d+1}, done\}$ from $\mathcal{B}$;
    Set the target Q-value for each experience
    $y^d = z^d + \gamma Q^-(s^{d+1}, \text{argmax}_a Q(s^{d+1}, a^d; \omega); \omega^-)$;
    Update ω by performing a gradient descent step and minimizing the loss function
    $L(\omega) = \mathbb{E}_{\{s^d, a^d, z^d, s^{d+1}, done\} \backsim \mathcal{B}(D)}[(y^d - Q(s^d, a^d; \omega))^2]$;
    Set $\omega^- \leftarrow \omega$ via a soft-updating method to update target network;
    if *done* **then** break.

the information on the available time slot resources of the considered TSN can be obtained for scheduling the next traffic. In addition, done $\in$ {True, False} label is utilized to indicate whether the time slot resources are exhausted up to now. The corresponding experience tuple is stored in replay buffer $\mathcal{B}$ for algorithm training, and it will replace the oldest tuple when the replay buffer is full.

(3) *Neural Network Training*: As previously described, the online and target networks are trained using the sampled experiences. In each algorithm step, the learning agent randomly selects a minibatch of D_b experiences from the replay buffer and updates the parameters of both networks.

Firstly, to update the online network, the loss function can be expressed by

$$L(\omega) = \mathbb{E}_{\{s^d, a^d, z^d, s^{d+1}\} \backsim \mathcal{B}(D)} \left[(y^d - Q(s^d, a^d; \omega))^2 \right], \tag{4.23}$$

where

$$y^d = z^d + \gamma Q^- \left(s^{d+1}, \text{argmax}_a Q(s^{d+1}, a^d; \omega); \omega^- \right), \tag{4.24}$$

is the target Q-value. Here, if done = True, then $y^d = z^d$. Then, parameter ω of the online network can be trained step-by-step via minimizing the loss function, i.e.,

$$\nabla_{\omega} L(\omega) = \mathbb{E}_{\{s^d, a^d, z^d, s^{d+1}\} \backsim \mathcal{B}(D)} [(y^d - Q(s^d, a^d; \omega)) \cdot \nabla_{\omega} Q(s^d, a^d; \omega)]. \tag{4.25}$$

Secondly, to guarantee the stability of the training process, a soft-updating method is adopted, and the parameters ω^- of the target network are slowly updated by tracking the online network, namely,

$$\omega^- = \kappa\omega + (1 - \kappa)\omega^-, \tag{4.26}$$

where $\kappa \in (0, 1]$ is the update ratio of the target network.

Thirdly, the DynamicSched adopts the $\varepsilon - greedy$ policy to keep balance between exploitation and exploration for making accurate decision a^t, in which the possibility factor ε decreases after each learning step, i.e., $\varepsilon = \varepsilon_0 - N$.

Remark 1: Since the number of input neurons of the agent is related to the devices' number, the network topology and device number cannot change during the training process. If new devices apply to participate in the DetFed framework, the number of input neurons of the agent should be adjusted accordingly, and the proposed DynamicSched algorithm needs to be retrained according to the new network topology. After a certain number of iterations, the updated scheduling decisions are obtained and can be deployed in the central controller to support DetFed operation.

Computation Complexity Analysis: As shown in Fig. 4.3, the DynamicSched algorithm is composed of an online network and a target network, and both of them are instantiated by an FC neural network. The computation complexity of the proposed resource scheduling algorithm is mainly correlated to the FC neural networks and the learning process. The computation complexity of an FC neural network is generally denoted by $\mathcal{O}(\sum_l (2M_{l-1} - 1)M_l)$ [50], where $l \in [1, N]$ is the index of hidden layers and M_l is the corresponding neuron number. Notably, taking into account the dueling structure is adopted in the DynamicSched algorithm, the output layer consists of two parts, i.e., state value function and state-dependent decision advantage function, and hence $M_N = M_N^{\text{state}} + M_N^{\text{decision}}$. Here, M_N^{state} and M_N^{decision} are the dimensions of the above two functions. For the learning process, the computation complexity is correlated to the step number $|K|$ and episode number $|E_{\text{num}}|$. To sum up, the total computation complexity of the DynamicSched can be denoted as $|K| \times |E_{\text{num}}| \times 2\mathcal{O}(\cdot)$.

The *relationship* between Algorithms 1 and 2 in Chap. 3 is as follows. Overall, both algorithms are utilized to support the proposed DetFed framework, and Algorithm 1 in Chap. 3 requires the scheduling decision made by Algorithm 2 in Chap. 3 in each FL round. Specifically, Algorithm 1 in Chap. 3 is to accelerate collaborative DNN model training process, which is operated round-by-round. Algorithm 2 in Chap. 3 is proposed to make network resource allocation decisions, which is operated episode-by-episode. In each episode, a learning agent deployed in a central controller acquires network states and makes decisions to support deterministic model parameter transmission during the operation of Algorithm 1 in Chap. 3.

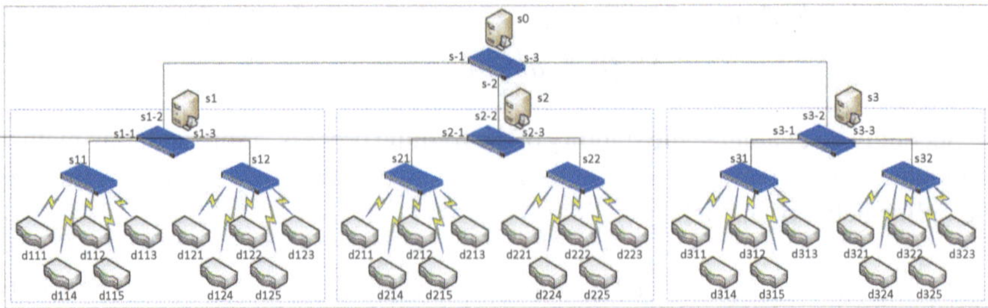

Fig. 4.4 The considered network topology in the experiment

4.5 Performance Evaluation and Analysis

In this part, extensive simulations are carried out to evaluate the proposed solutions for accelerating and optimizing FL over the wireless and wired integrated TSN.

The network topology is shown as Fig. 4.4, where 30 devices, 3 field servers, and a factory server are considered to support DetFed operation which are connected as a tree topology by 10 TSN switches. Thus, there are 66 FL traffic needs to be scheduled within one scheduling cycle (i.e., one FL round).[1] The link bandwidth for the wireless and wired integrated TSN is set to 1 Gbit/s unless specified. Note that 90% of the bandwidth is utilized to support DetFed operation, and 10% of the bandwidth is reserved to transmit other industrial traffic, such as control- and/or safety-critical traffic. The propagating delay τ_{prop}, processing delay τ_{proc}, queuing delay τ_{queue}, and synchronizing delay τ_{sync} are set to 10 μs. The traffic period $K[\text{periods}]$ is set to 40 ms, and thus the scheduling cycle SC is 40 ms. The size of time slots within each scheduling cycle is set to 10 ms. For the DynamicSched algorithm, the online and target networks are with [S_{dim}, 1000, 500, A_{dim}] neurons, respectively. Here, S_{dim} is the dimension of the observed environment state, and A_{dim} is the time slot number of the considered network scenario. The update factor τ of target network is set to 0.005, which is decided by the scheduling cycle SC and the time slot size. The reward factor V_{succ} and penalty factor V_{fail} are set to be 0.1 and $-$ 0.1, respectively. Other important simulation parameters are listed in Table 4.2.

We consider an attention-enhanced convolutional recurrent neural network (ACRNN), which contains 10 layers, i.e., two convolutional (Conv) layers, two Max-pool layers, a flatten layer, an FC layer, a forward gated recurrent unit (GRU) layer, a backward GRU layer, an attention layer, and a softmax output layer. The detailed FL model information is shown in Table 4.3. The total trainable model parameters are around 15,099. Here, each parameter is quantized to 4 B, and thus the data size of the model parameters $K[\text{sizes}]$ is around 58.98 kB.

[1] For model parameter distribution, three FL traffic are distributed from the factory server to the field servers, and thirty FL traffic are distributed from the field servers to the devices. For model parameter uploading, FL traffic is uploaded in the opposite direction, and the number of FL traffic is the same.

Table 4.2 Simulation parameters [44, 48]

DetFed param.	Value	DynamicSched param.	Value
Optimizer	Adam	Optimizer$^-$	Adam
Round	300	$\|E_{\text{num}}\|$	800
S_{size}	23,436	$\|K\|$	66
Batch	64	Batch$^-$	128
$(\text{AWE}_{\text{in}}, \text{AWE}_{\text{out}})$	(1024, 6)	Buffer	10,000
$(\text{CWRU}_{\text{in}}, \text{CWRU}_{\text{out}})$	(4096, 10)	γ	0.95
lr	2.5×10^{-3}	lr$^-$	1×10^{-4}
λ	8×10^{-5}	ε	0.5

Table 4.3 ACRNN structure [49]

Layer name	Kernal size	Activation func.	Param.#
Conv-1	(5, 5, 1, 6)	ReLU	156
Maxpool-1	(1, 2, 2, 1)	–	12
Conv-2	(5, 5, 6, 16)	ReLU	1516
Maxpool-2	(1, 2, 2, 1)	–	32
Flatten	(400, 1)	–	–
FC	(400, 16)	tanh	6400
Forward GRU	(32, 1)	tanh/sigmoid	3264
Backward GRU	(32, 1)	tanh/sigmoid	3264
Attention	(64, 1)	tanh	65
Output	(64, 6)	Softmax	390

In the simulation, we utilize two vibration-based datasets to train the ACRNN model: (1) *AWE dataset*[2] in which each data sample is a vibration signal segment attached with a label to represent the health status of an industrial equipment, such as "Normal" and "Fault-*6mm*"; and (2) *CWRU dataset*[3] in which each data sample is also a vibration signal segment attached with a label to indicate the fault location, such as "Inner Race" and "Outer Race". In both datasets, the number of data samples are 23,436 and 479, respectively, and the ratio of training set to testing set is 9:1. Notably, the considered ACRNN model should be trained on a non-independent and identical distribution (non-IID) dataset, which is dispersed at the distributed participating devices and each device has only a few classes of data samples. Specifically, the samples of both datasets are sorted by their data label, respectively, which are further divided into 60 subsets with the same sample volume. Then, each device randomly selects 2 subsets as its local training set.

[2] Online available. https://github.com/Intelligent-AWE/DeepHealth.

[3] Online available. https://csegroups.case.edu/bearingdatacenter/pages/download-data-file.

The proposed solutions are compared to the following benchmark schemes.

- ***Centralized Training***: This scheme requires uploading all the dispersed data to the factory server to conduct model training. Although it can achieve the optimal learning accuracy, the data privacy cannot be preserved effectively.
- ***The DetFed Without Field Aggregation (DetFed w.o.)***: In this scheme, all the updated model parameters are only aggregated by the factory server, which occupies more bandwidth resources of industrial buses.
- ***DDQN***: In this scheme, the decisions are dynamically made via a DDQN algorithm, in which the Q-network is endowed with two fully-connected hidden layers and each layer is with 1000 neurons. Other learning parameters are the same as the DynamicSched.
- ***Random-Based Scheme (RS)***: This scheme adopts random policy to allocate temporal resources for each FL traffic, and all available decisions are made with an equal probability.

Figure 4.5a shows the convergence performance of the proposed DetFed scheme with respect to different learning rates for the ACRNN-based model. When the learning rate $\mathrm{lr} = 0.0025$, the proposed scheme achieves the highest testing accuracy and the lowest testing loss, respectively, although it tends to decline around 60 rounds before converging to the optimal accuracy. In addition, it can be seen that a larger learning rate (e.g., $\mathrm{lr} = 0.005$) may incur the training process because the optimal value of accuracy might be skipped, and thus leading to an inferior convergence performance. Although a small learning rate helps stabilizing the training process, the ACRNN-based model may be trapped in a local optimum, and consequently may need to consume more temporal resources to support model parameter transmission and achieve model convergence. A similar simulation is carried out on the CWRU dataset. As shown in Fig. 4.5b, the highest testing accuracy and the lowest testing loss can be obtained when the learning rate is 0.005.

Figure 4.6 shows the convergence performance of the proposed DetFed scheme with respect to different factory aggregation intervals. In the DetFed framework, the factory aggregation is performed every r_g FL rounds, in which the parameters aggregated at the field servers are further aggregated at a factory server. We can obtain several important observations from the simulation results. Firstly, when the factory aggregation interval r_g is set to 16, the highest accuracy on the AWE dataset can be obtained despite an inferior training performance is exhibited before 40 rounds. Secondly, if we set the factory aggregation interval r_g as an appropriate value, the training process can be effectively accelerated. Specifically, the proposed scheme with $r_g = 16$ achieves the fastest convergence after 170 training rounds, which decreases convergence round number by 16 and 50 compared with $r_g = 8$ and $r_g = 24$, respectively. Thirdly, the proposed DetFed scheme achieves almost the same testing accuracy at the end of the training process. This indicates that the TSN-enhanced DetFed scheme can provide efficient and deterministic transmission services for model parameter aggregation and distribution among the three layers.

As shown in Fig. 4.7, we evaluate the testing accuracy of the proposed DetFed scheme and benchmarks in terms of different parameter aggregation locations. The

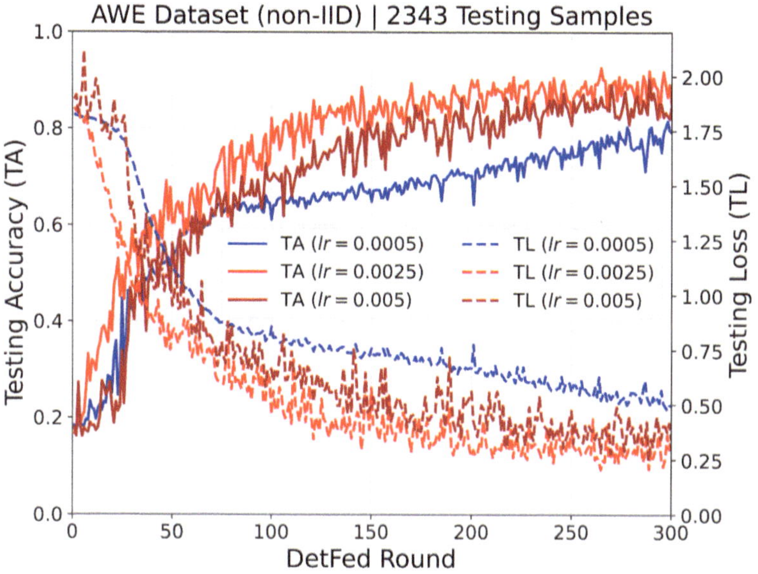

(a) AWE dataset

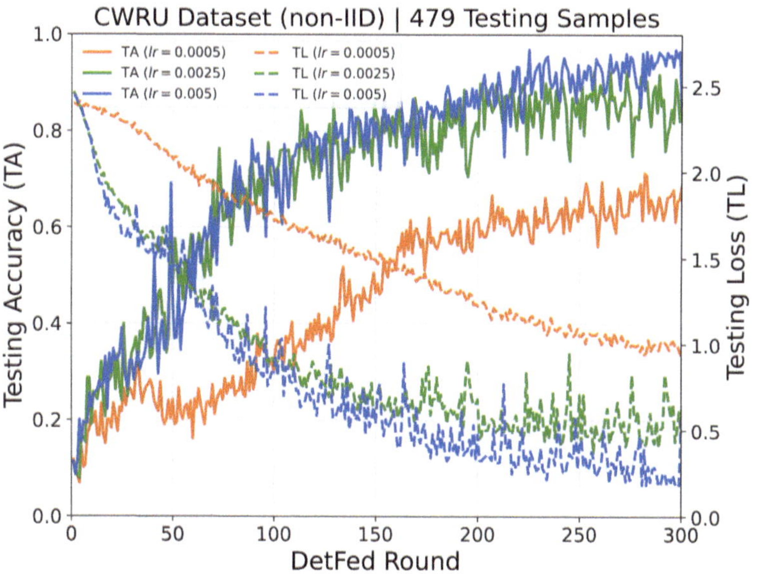

(b) CWRU dataset

Fig. 4.5 Testing accuracy and loss of DetFed with respect to different learning rates

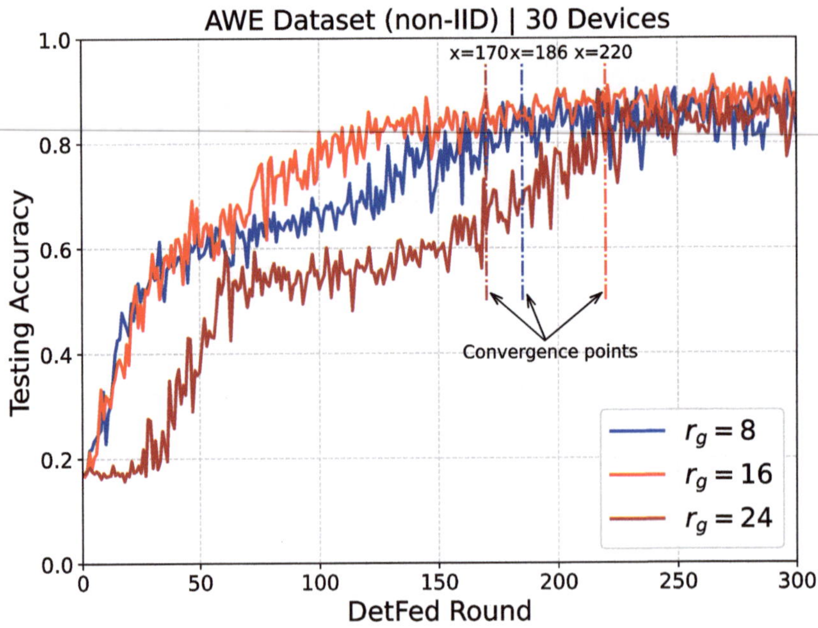

Fig. 4.6 Testing accuracy of DetFed with respect to different factory aggregation frequencies

result shows that all the schemes can achieve model convergence within 300 training rounds. Intuitively, the CT scheme only takes a few training rounds to converge (around 50 rounds) and can converge to an especially high level (i.e., > 99%). This is because a centralized dataset has been stored at the factory server that were collected from the dispersed IIoT devices before training. Such a scheme indeed can obtain a satisfactory convergence performance without parameter aggregation demand, but it inevitably exists the problems of data transmission overhead and data privacy leakage which are critical issues in industrial IoT scenarios. Although the distributed learning schemes, namely DetFed and DetFed w.o., only converge to a relatively low level (i.e., $\approx$ 90%), they can effectively save bandwidth resources of industrial buses and preserve privacy for industrial raw data. In particular, the DetFed scheme can reduce the number of training rounds until model convergence by 29.17% compared with the DetFed w.o. scheme. The reason is that directly aggregating model parameters at the factory server may lead to an unstable learning process due to the non-IID data distribution, which impedes the ACRNN model to obtain the global optimal parameters and slows down the convergence speed. On the contrary, the proposed DetFed scheme stabilizes the learning process by alternately aggregating model parameters among a factory server and field servers, such that the global optimal parameters can be efficiently obtained based on local optimal parameters and thus accelerating model convergence.

As shown in Figs. 4.6 and 4.7, the proposed DetFed framework can effectively accelerate the collaborative training process in the considered network topology.

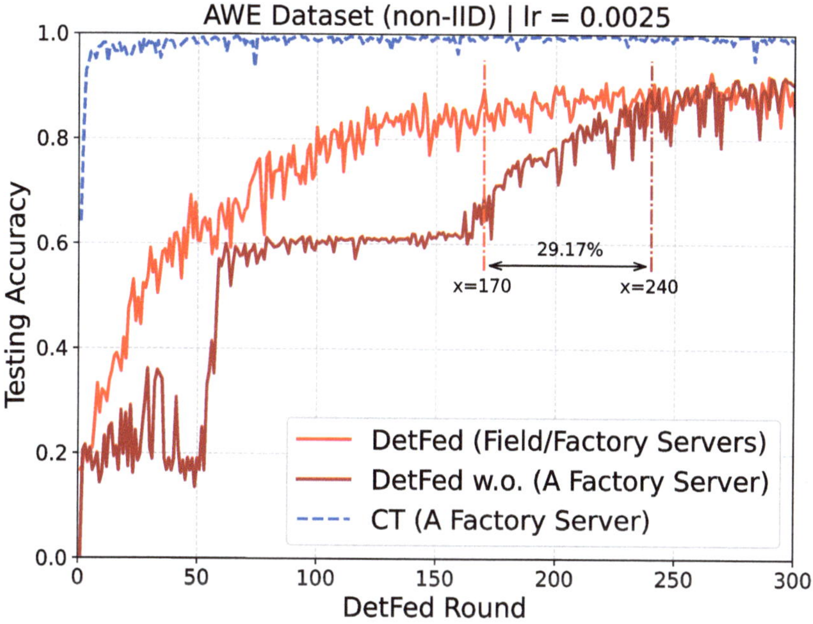

Fig. 4.7 Testing accuracy comparison among the proposed scheme and benchmarks with respect to different parameter aggregation locations

However, this result is achieved under the assumption that the IoT connections are stable. In harsh network scenarios, where serious interference and electrical noises exist, and the network bandwidth resources utilization is not high, the transmission delay for FL model parameters may not be satisfied. In particular, when some IoT devices are disconnected, the updated model parameters will not be uploaded to the aggregation servers, which further slows down the convergence rate.

As shown in Fig. 4.8, the convergence performance of the DynamicSched with respect to episodes is evaluated. It can be seen that all the learning-based resource scheduling algorithms have converged. In particular, the DynamicSched achieves a higher episode reward compared with other benchmarks, which indicates that the proposed DynamicSched is capable of making judicious allocation decisions for temporal resources of the considered TSN. This is because the DynamicSched adopts a D3QN-based architecture, in which an online-target *double* module and a value-advantage *dueling* module are utilized to improve algorithm performance. Specifically, the dueling module is helpful to capture which network states are (or are not) useful without having to learn the effect of each resource allocation decision for each network state, and thus identifying the appropriate decision more judiciously.

Figure 4.9 shows average scheduling success ratio with respect to different bandwidth resources over 100 simulation runs. The result shows that the proposed algorithm can effectively improve average scheduling success ratio compared with the benchmark algorithms, which indicates that temporal resource allocation is opti-

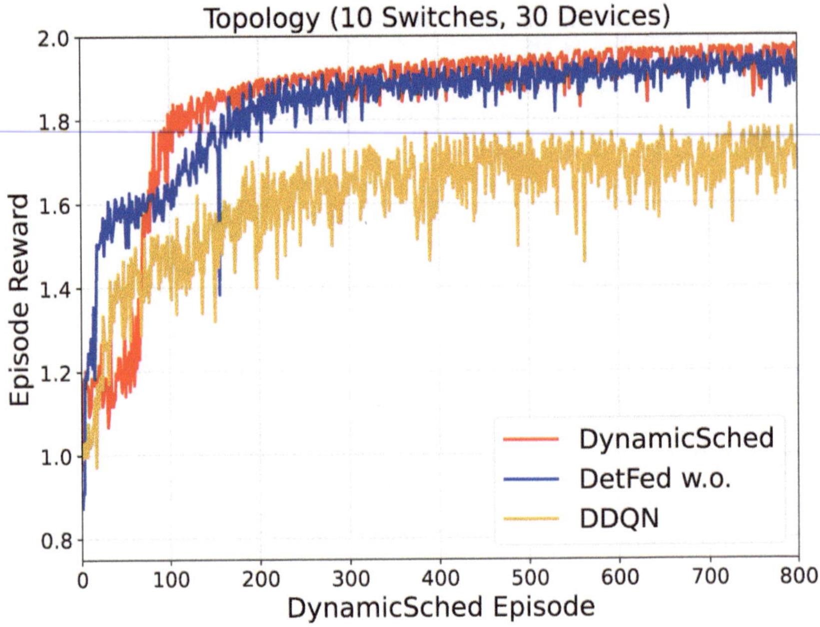

Fig. 4.8 Convergence performance comparison among different algorithms

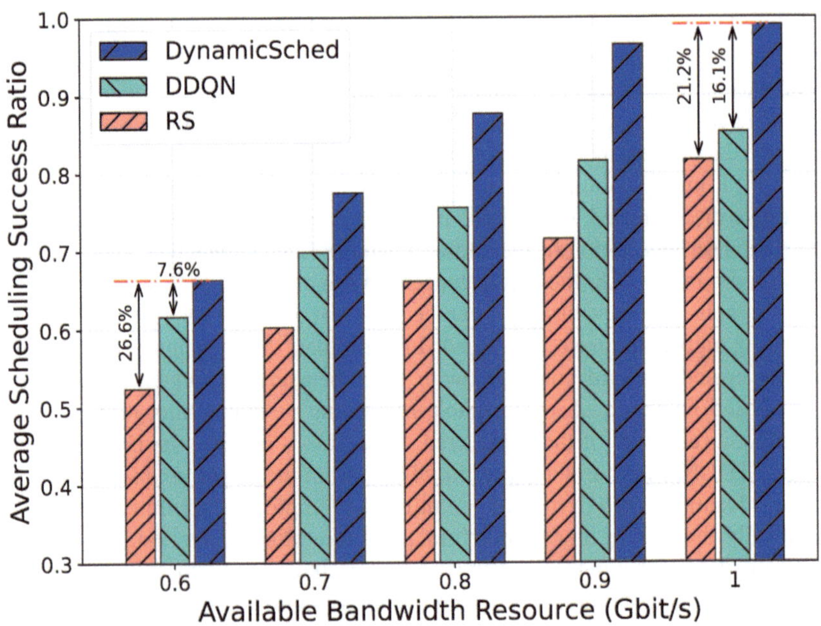

Fig. 4.9 Average scheduling success ratio comparison among the proposed scheme and benchmarks with respect to different bandwidth resources

mized. In specific, when the amount of bandwidth resource is 1 Gbit/s, the proposed DynamicSched improves the average scheduling success ratio by 16.1% and 21.2% compared with the DDQN and RS benchmarks, respectively. This is because the proposed learning-based algorithm can allocate temporal resources for FL traffic more reasonably, which is helpful to satisfy the deterministic transmission requirements of FL traffic. In addition, the performance gain achieved in bandwidth-sufficient scenarios (e.g., 1 Gbit/s) is larger than that in bandwidth-limited scenarios (e.g., 0.6 Gbit/s), as compared to the DDQN benchmark. The result indicates that integrating the dueling module into the double module can enhance the decision-making capability of the proposed algorithm for temporal resource allocation, thereby improving the scheduling success ratio, especially when the bandwidth resource is sufficient.

Figure 4.10 shows the average scheduling success ratio in terms of different bandwidth reservation ratios. Taking the value of 0.9 as an example, when the bandwidth reservation ratio is set to 0.9, 90% of bandwidth resources are utilized to support DetFed operation and 10% of bandwidth resources are reserved to support the co-transmission of burst traffic. As expected, the DynamicSched algorithm can effectively improve the average scheduling success ratio as compared with the benchmarks. This is because the proposed algorithm can better allocate bandwidth resources to support FL traffic transmission with satisfied deterministic requirements of delay, jitter, and packet loss. In addition, with the increase of the bandwidth reservation ratios, the average scheduling success ratios achieved by the DynamicSched,

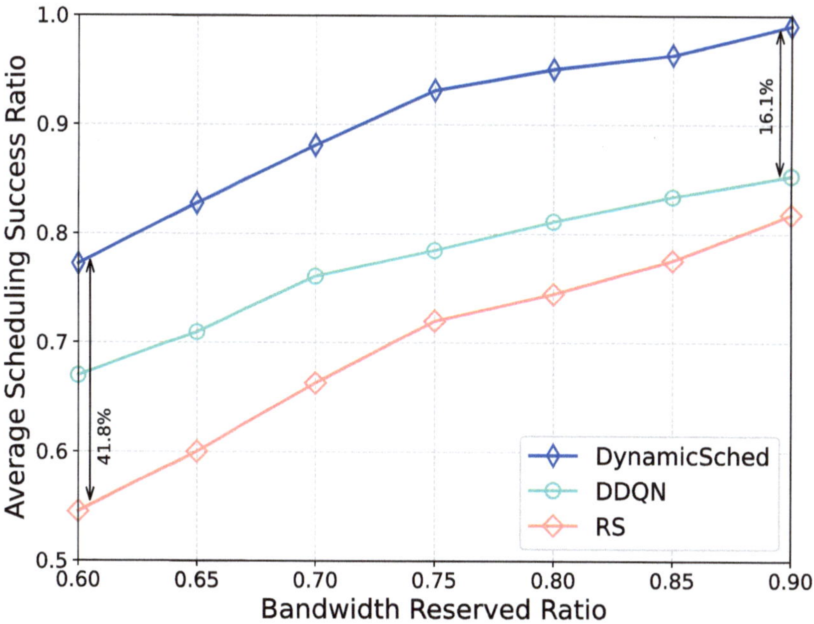

Fig. 4.10 Average scheduling success ratio comparison among the proposed scheme and benchmarks with respect to different bandwidth reservation ratios

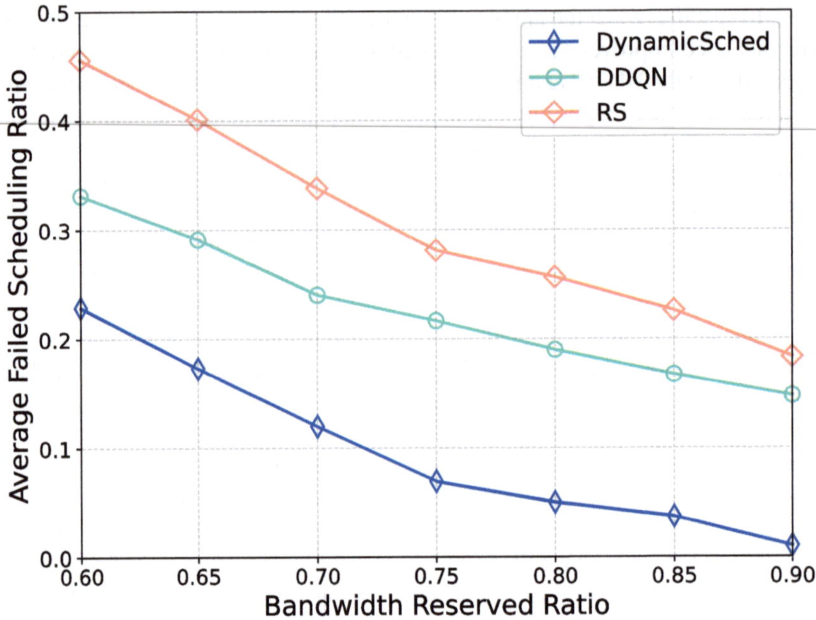

Fig. 4.11 Average failed scheduling ratio comparison among the proposed scheme and benchmarks with respect to different bandwidth reservation ratios

DDQN, and RS algorithms increase. In particular, the proposed algorithm improves the average scheduling success ratio by 15.3% to 16.1%, and by 41.8% to 21.2%, as compared with the DDQN and RS benchmarks, respectively. The underlying reason is that more available bandwidth resources can be utilized to determinately transmit FL traffic over the considered TSN scenario. To demonstrate the algorithm performance in the case of failed scheduling, we have presented the average failed scheduling ratio of the proposed DynamicSched algorithm with respect to bandwidth reserved ratios. As shown in Fig. 4.11, the DynamicSched achieves the lowest average failed scheduling ratio compared to all the benchmarks. In addition, the failed scheduling ratio decreases with the amounts of reserved bandwidth resources. The result indicates that the DynamicSched can achieve good performance due to effective resource utilization.

Figure 4.12 shows the impact of the time slot sizes on scheduling performance. The result shows that the learning-based algorithms, i.e., DynamicSched and DDQN, can achieve higher average scheduling success ratio than that of RS benchmark. The reason is that learning-based algorithms enable the deterministic requirements of FL traffic to be satisfied efficiently via judicious temporal resource scheduling. Moreover, with the increase of the time slot sizes, the average scheduling success ratio of the compared algorithms increase. This is because that more available temporal resources can be utilized to schedule FL traffic to satisfy their deterministic delay, jitter, and packet loss requirements. In particular, the proposed DynamicSched algo-

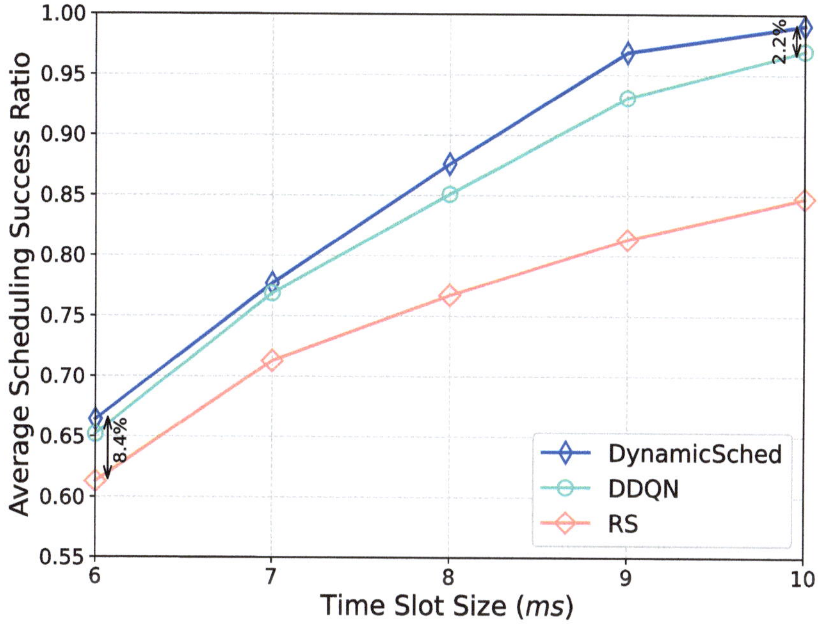

Fig. 4.12 Average scheduling success ratio comparison among the proposed scheme and benchmarks with respect to different time slot sizes

rithm can achieve the highest average scheduling success ratio than the benchmarks. When the size of time slot is 10 ms, the DynamicSched can increase the average scheduling success ratio by 2.2% and 16.9%, as compared with DDQN and RS benchmarks, respectively.

4.6 Summary

We have presented a new deterministic FL framework (i.e., DetFed) for the time-sensitive IIoT in combination with 6G-oriented ultra-reliable and low-latency communication techniques. We have proposed a DRL-based resource scheduling algorithm to make the optimal temporal resource allocation decisions for facilitating efficient DetFed, which aggregates more device-side models for obtaining an accurate global model. Experimental results based on real-world non-IID dataset have demonstrated that the proposed framework can significantly accelerate FL convergence and improve learning accuracy as compared with state-of-the-art benchmarks. By adopting the wireless and wired integrated TSN and the CQF mechanism, the DetFed can optimize the learning accuracy of FL without affecting the co-transmission of burst traffic. For the future work, we will study a mixed and online traffic scheduling scheme for facilitating efficient FL over a large-scale and distributed computing power network.

References

1. H. Kang, Z. Li, Q. Zhang, Communicational and computational efficient federated domain adaptation. IEEE Trans. Parallel Distrib. Syst. **33**(12), 3678–3689 (2022)
2. X. Shen, J. Gao, W. Wu, K. Lyu, M. Li, W. Zhuang, X. Li, J. Rao, AI-assisted network-slicing based next-generation wireless networks. IEEE Open J. Veh. Technol. **1**(1), 45–66 (2020)
3. B. Yang, X. Cao, J. Bassey, X. Li, L. Qian, Computation offloading in multi-access edge computing: a multi-task learning approach. IEEE Trans. Mob. Comput. **20**(9), 2745–2762 (2021)
4. L.U. Khan, W. Saad, Z. Han, E. Hossain, C.S. Hong, Federated learning for internet of things: recent advances, taxonomy, and open challenges. IEEE Commun. Surv. Tutor. **23**(3), 1759–1799 (2021)
5. C. Zhang, M. Zhao, L. Zhu, W. Zhang, T. Wu, J. Ni, FRUIT: a blockchain-based efficient and privacy-preserving quality-aware incentive scheme. IEEE J. Sel. Areas Commun. **40**(12), 3343–3357 (2022)
6. J. Kang, X. Li, J. Nie, Y. Liu, M. Xu, Z. Xiong, D. Niyato, Q. Yan, Communication-efficient and cross-chain empowered federated learning for artificial intelligence of things. IEEE Trans. Netw. Sci. Eng. **9**(5), 2966–2977 (2022)
7. Z. Qu, S. Guo, H. Wang, B. Ye, Y. Wang, A.Y. Zomaya, B. Tang, Partial synchronization to accelerate federated learning over relay-assisted edge networks. IEEE Trans. Mob. Comput. **21**(12), 4502–4516 (2022)
8. W.Y.B. Lim, Z. Xiong, C. Miao, D. Niyato, Q. Yang, C. Leung, H.V. Poor, Hierarchical incentive mechanism design for federated machine learning in mobile networks. IEEE Internet Things J. **7**(10), 9575–9588 (2020)
9. F. Sattler, S. Wiedemann, K.-R. Müller, W. Samek, Robust and communication-efficient federated learning from non-i.i.d. data. IEEE Trans. Neural Netw. Learn. Syst. **31**(9), 3400–3413 (2020)
10. Q. Tang, F.R. Yu, R. Xie, A. Boukerche, T. Huang, Y. Liu, Internet of intelligence: a survey on the enabling technologies, applications, and challenges. IEEE Commun. Surv. Tutor. **24**(3), 1394–1434 (2022)
11. E. Li, F. He, Q. Li, H.G. Xiong, Bandwidth allocation of stream-reservation traffic in TSN. IEEE Trans. Netw. Serv. Manag. **19**(1), 741–755 (2022)
12. M. Chen, D. Gündüz, K. Huang, W. Saad, M. Bennis, A.V. Feljan, H.V. Poor, Distributed learning in wireless networks: recent progress and future challenges. IEEE J. Sel. Areas Commun. **39**(12), 3579–3605 (2021)
13. J. Prados-Garzon, T. Taleb, M. Bagaa, Optimization of flow allocation in asynchronous deterministic 5G transport networks by leveraging data analytics. IEEE Trans. Mob. Comput. **23**(3), 1672–1687 (2023)
14. M. Vlk, Z. Hanzálek, K. Brejchová, S. Tang, S. Bhattacharjee, S. Fu, Enhancing schedulability and throughput of time-triggered traffic in IEEE 802.1Qbv time-sensitive networks. IEEE Trans. Commun. **68**(11), 7023–7038 (2020)
15. J. Prados-Garzon, T. Taleb, M. Bagaa, LEARNET: reinforcement learning based flow scheduling for asynchronous deterministic networks, in *Proceedings of IEEE ICC 2020*, Dublin, Ireland (2020), pp. 1–6
16. Z. Pang, X. Huang, Z. Li, S. Zhang, Y. Xu, H. Wan, X. Zhao, Flow scheduling for conflict-free network updates in time-sensitive software-defined networks. IEEE Trans. Ind. Inform. **17**(3), 1668–1678 (2021)
17. Y. Huang, S. Wang, B. Wu, T. Huang, Y. Liu, TACQ: enabling zero-jitter for cyclic-queuing and forwarding in time-sensitive networks, in *Proceedings of IEEE ICC*, Montreal, QC, Canada (2021), pp. 1–6
18. D. Yang, Z. Cheng, W. Zhang, H. Zhang, X. Shen, Burst-aware time-triggered flow scheduling with enhanced multi-CQF in time-sensitive networks. IEEE/ACM Trans. Netw. 31(6), 2809–2824 (2023)

19. X. Shen, J. Gao, W. Wu, M. Li, C. Zhou, W. Zhuang, Holistic network virtualization and pervasive network intelligence for 6G. IEEE Commun. Surv. Tutor. **24**(1), 1–30 (2021)
20. M.N. Nguyen, N.H. Tran, Y.K. Tun, Z. Han, C.S. Hong, Toward multiple federated learning services resource sharing in mobile edge networks. IEEE Trans. Mob. Comput. **22**(1), 541–555 (2023)
21. W. Zhang, D. Yang, W. Wu, H. Peng, N. Zhang, H. Zhang, X. Shen, Optimizing federated learning in distributed industrial IoT: a multi-agent approach. IEEE J. Sel. Areas Commun. **39**(12), 3688–3703 (2021)
22. M. Chen, Z. Yang, W. Saad, C. Yin, H.V. Poor, S. Cui, A joint learning and communications framework for federated learning over wireless networks. IEEE Trans. Wireless Commun. **20**(1), 269–283 (2021)
23. S. Wan, J. Lu, P. Fan, Y. Shao, C. Peng, K.B. Letaief, Convergence analysis and system design for federated learning over wireless networks. IEEE J. Sel. Areas Commun. **39**(12), 3622–3639 (2021)
24. C.T. Dinh, N.H. Tran, M.N.H. Nguyen, C.S. Hong, W. Bao, A.Y. Zomaya, V. Gramoli, Federated learning over wireless networks: convergence analysis and resource allocation. IEEE/ACM Trans. Netw. **29**(1), 398–409 (2021)
25. M. Salehi, E. Hossain, Federated learning in unreliable and resource-constrained cellular wireless networks. IEEE Trans. Commun. **69**(8), 5136–5151 (2021)
26. J. Zhang, N. Li, M. Dedeoglu, Federated learning over wireless networks: a band-limited coordinated descent approach, in *Proceedings of IEEE INFOCOM*, Vancouver, BC, Canada (2021), pp. 1–10
27. D. Yang, K. Gong, J. Ren, W. Zhang, W. Wu, H. Zhang, TC-flow: chain flow scheduling for advanced industrial applications in time-sensitive networks. IEEE Netw. **36**(2), 16–24 (2022)
28. X. Lu, Q. Xu, X. Wang, M. Lin, C. Chen, Z. Shi, X. Guan, Full-loop AoI-based joint design of control and deterministic transmission for industrial CPS. IEEE Trans. Ind. Inform. **19**(11), 10727–10738 (2023)
29. C. Zou, H. Zhao, L. Geng, Q. Zhao, D. Wang, Optimization for efficient federated learning: joint scheduling in vehicular edge computing networks, in *Proceedings of IEEE GLOBECOM* (2024), pp. 3944–3949
30. Z. Ni, Z. Zhang, N.C. Luong, D. Niyato, D.I. Kim, S. Feng, Joint client scheduling and quantization optimization in energy harvesting-enabled federated learning networks. IEEE Trans. Wireless Commun. **23**(8), 9566–9582 (2024)
31. C. Wang, Y. Peng, J. Wu, L. Liu, S. Mumtaz, M. Dong, M. Guizani, Deterministic scheduling and reliable routing for smart ocean services in maritime internet of things: a cross-layer approach. IEEE Trans. Serv. Comput. **17**(6), 3387–3399 (2024)
32. X. Chen, Z. Li, W. Ni, X. Wang, S. Zhang, Y. Sun, S. Xu, Q. Pei, Toward dynamic resource allocation and client scheduling in hierarchical federated learning: a two-phase deep reinforcement learning approach. IEEE Trans. Commun. **72**(12), 7798–7813 (2024)
33. W. Wang, W. Wang, H. Yu, B. Wu, S. Zou, W. Ni, J. Zhang, Joint routing and scheduling for in-vehicle networks: a deterministic perspective. IEEE Trans. Intell. Veh. 1–15 (2024)
34. G. Li, S. Wang, Y. Huang, T. Huang, Y. Cui, Z. Xiong, Optimizing fault-tolerant time-aware flow scheduling in TSN-5G networks. IEEE Trans. Mob. Comput. **24**(4), 3441–3455 (2025)
35. G. Yao, A.M. Bedewy, N.B. Shroff, Age-optimal low-power status update over time-correlated fading channel. IEEE Trans. Mob. Comput. **22**(8), 4500–4514 (2022)
36. L. Zhang, G. Sun, R. Liu, W. Quan, H. Yu, D. Niyato, Priority-dominated traffic scheduling enabled ATS in time-sensitive networking. IEEE Trans. Netw. Serv. Manag. **22**(1), 470–484 (2025)
37. A.A. Atallah, G.B. Hamad, O.A. Mohamed, Routing and scheduling of time-triggered traffic in time-sensitive networks. IEEE Trans. Ind. Inform. **16**(7), 4525–4534 (2020)
38. J. Yan, W. Quan, X. Jiang, Z. Sun, Injection time planning: making CQF practical in time-sensitive networking, in *Proceedings of IEEE INFOCOM*, Beijing, China (2020), pp. 616–625
39. L. Zhao, P. Pop, Z. Zheng, H. Daigmorte, M. Boyer, Latency analysis of multiple classes of AVB traffic in TSN with standard credit behavior using network calculus. IEEE Trans. Ind. Electron. **68**(10), 10291–10302 (2021)

40. Y. Yuan, X. Cao, Z. Liu, C. Chen, X. Guan, Adaptive priority adjustment scheduling approach with response time analysis in time-sensitive networks. IEEE Trans. Ind. Inform. **18**(12), 8714–8723 (2022)
41. Y. Zhang, Q. Xu, X. Guan, C. Chen, M. Li, Wireless/wired integrated transmission for industrial cyber-physical systems: risk-sensitive co-design of 5G and TSN protocols. Sci. China Inf. Sci. **65**(1), 1–16 (2021)
42. Z. Cheng, D. Yang, W. Zhang, J. Ren, H. Wang, H. Zhang, DeepCQF: making CQF scheduling more intelligent and practicable, in *Proceedings of IEEE ICC*, Seoul, Korea (2022), pp. 1–6
43. IEEE standard for local and metropolitan area networks—bridges and bridged networks—amendment 31: stream reservation protocol (SRP) enhancements and performance improvements, in *IEEE Std 802.1Qcc-2018 (Amendment to IEEE Std 802.1Q-2018 as amended by IEEE Std 802.1Qcp-2018)* (2018), pp. 1–208
44. H. Sami, H. Otrok, J. Bentahar, A. Mourad, AI-based resource provisioning of IoE services in 6G: a deep reinforcement learning approach. IEEE Trans. Netw. Serv. Manag. **18**(3), 3527–3540 (2021)
45. Q. Ye, W. Shi, K. Qu, H. He, W. Zhuang, X. Shen, Joint RAN slicing and computation offloading for autonomous vehicular networks: a learning-assisted hierarchical approach. IEEE Open J. Veh. Technol. **2**, 272–288 (2021)
46. S. Wang, Y. Guo, N. Zhang, P. Yang, A. Zhou, X. Shen, Delay-aware microservice coordination in mobile edge computing: a reinforcement learning approach. IEEE Trans. Mob. Comput. **20**(3), 939–951 (2021)
47. Y. Wang, M. Chen, T. Luo, W. Saad, D. Niyato, H.V. Poor, S. Cui, Performance optimization for semantic communications: an attention-based reinforcement learning approach. IEEE J. Sel. Areas Commun. **40**(9), 2598–2613 (2022)
48. W. Shi, S. Zhou, Z. Niu, M. Jiang, L. Geng, Joint device scheduling and resource allocation for latency constrained wireless federated learning. IEEE Trans. Wireless Commun. **20**(1), 453–467 (2021)
49. W. Zhang, D. Yang, H. Wang, X. Huang, M. Gidlund, CarNet: a dual correlation method for health perception of rotating machinery. IEEE Sens. J. **19**(16), 7095–7106 (2019)
50. R. Livni, S. Shalev, O. Shamir, On the computational efficiency of training neural networks, in *Proceedings of NeurIPS*, vol. 27, Montréal, Canada (2014), pp. 855–863

Chapter 5
End-Edge-Cloud Orchestration for Collaborative DNN Inference

5.1 Introduction

With the advances in AI technologies [1], modern industry is developing towards smart automation. Emerging industrial Internet of things applications powered by AI, such as intelligent manufacturing, predictive maintenance, and automatic warehousing, are being extensively applied in practical industrial systems [2, 3]. Particularly, DNNs have become an essential component for these intelligent IIoT applications [4, 5]. For automated production lines (APLs), the operating status of industrial facilities, e.g., numerically-controlled machine tools, robot arms, and automated guided vehicles, needs to be monitored in real time [6]. Based on the collected sensing data from on-site IIoT devices, DNNs can efficiently perform the condition monitoring and predict the impending failures of industrial facilities, namely *DNN inference*, which can significantly improve the production efficiency [7–14].

As shown in Fig. 5.1, implementing DNN inference requires two stages, namely, offline training and online inference [15]. In the offline training stage, the parameters of the DNN models are updated via iteratively training, and the well-trained DNN models are stored in a cloud server. In the online inference stage, the pre-trained DNN models can be placed at edge servers or end devices, to perform the inference tasks. With the wide deployment of intelligent IIoT applications, a large amount of inference tasks need to be processed instantly, which requires high demands on computing resources [16]. Due to the limited computation capabilities of IIoT devices, computation-intensive inference tasks can be offloaded to the remote cloud server [17], yet long-distance data transmission results in a high response latency [18]. To provide delay-sensitive computing services, multi-access edge computing [19], arises as a novel paradigm by deploying substantial computing resources at the network edge [20]. When the pre-trained DNN models are placed at the edge servers, it is no longer necessary to transmit tasks to the cloud, thus DNN inference tasks can be performed with a low latency.

© The Author(s), under exclusive license to Springer Nature Switzerland AG 2026

W. Zhang et al., *Intelligent Resource Scheduling in End-Edge-Cloud Networks*, Wireless Networks, https://doi.org/10.1007/978-3-032-07667-0_5

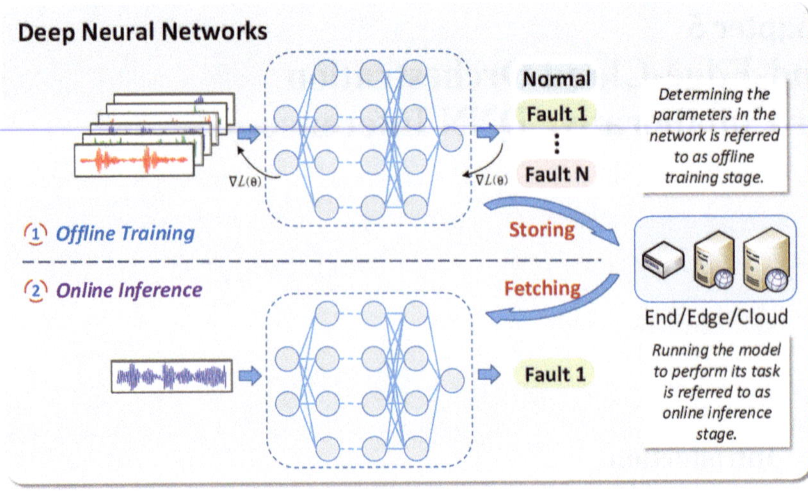

Fig. 5.1 The procedure of implementing DNN inference: (1) offline training and (2) online inference

Except for the computing resource, caching resource is another necessity for inference task execution. When an inference task is offloaded to a node, the task can be executed only if the required pre-trained DNN model is cached in this node. However, neither an edge server nor an end device has sufficient capacity to cache all the required DNN models. Therefore, how to flexibly place the pre-trained DNN models among network nodes with limited caching resources is of significance for supporting DNN inference. In addition, spectrum resource also has a big impact on the transmission delays of inference tasks and pre-trained DNN models, thus reasonable spectrum resource allocation can provide low-latency transmission for tasks and DNN models. Hence, how to jointly manage multi-dimensional resources in IIoT networks to deliver low-latency DNN inference is critical. Moreover, except for the delay requirements, inference accuracy becomes a particularly key performance metric for intelligent IIoT applications [1]. However, due to the heterogeneity of industrial facilities, dedicated pre-trained DNN models need to be provided for inference tasks based on the dynamic service requests, which poses a great challenge to efficiently manage multi-dimensional resources in IIoT networks to provide low-latency and high-accuracy inference services.

Recently, advanced learning-based algorithms [25–28] can be applied to improve the decision-making efficiency. Particularly, DRL approaches have been utilized to solve resource optimization problems in wireless networks [22]. Deep Q-learning (DQN) and policy gradient based algorithms [23, 24] have demonstrated the strong capability for solving the optimization problems with discrete and continuous decision variables, respectively. Moreover, unlike traditional optimization methods, DRL can effectively capture network dynamics while solving complicated optimization problems with a low time complexity, thereby improving the problem-solving efficiency. Hence, DRL is a promising method to address the multi-dimensional resource management problem for DNN inference.

In this chapter, aiming to maximize the average accuracy of inference tasks, we investigate how to leverage DRL to efficiently allocate multi-dimensional resources to ensure DNN inference can be completed within strict delay requirements. Specifically, an end-edge-cloud orchestration architecture is introduced to support massive inference tasks in IIoT networks. In addition, the pre-trained DNN models are flexibly placed at the end device and edge server to provide low-latency inference services. Then, a joint inference task assignment and DNN model placement problem is formulated to maximize the average inference accuracy under multi-dimensional resource constraints. To efficiently solve the optimization problem, we therefore transform the formulated problem into a Markov decision process, and then propose a twin delayed DDPG-based resource management scheme to obtain the solution rapidly. Extensive simulation results are provided to demonstrate that the proposed learning-based resource management scheme can improve the average accuracy of inference tasks while satisfying the strict delay requirements. The main contributions of this chapter are summarized as follows.

- We introduce an end-edge-cloud orchestration architecture to flexibly coordinate the inference task assignment and DNN model placement, thereby providing low-latency DNN inference services in IIoT networks.
- We formulate a joint task assignment and DNN placement problem to maximize the average accuracy of massive inference tasks while satisfying the delay requirements under constrained spectrum, caching, and computing resources.
- To effectively solve the joint optimization problem, we transform the TADP problem into an MDP, and propose a DRL-based algorithm to learn the inference task assignment and DNN model placement policy and make instant resource allocation decisions for DNN inference.

The remainder of this chapter is organized as follows. Section 5.2 presents related works. Section 5.3 describes the system model and problem formulation. In Sect. 5.4, the learning-based resource management scheme is proposed, including the transformation of the formulated problem. Section 5.5 presents the performance evaluations for the proposed algorithm, and the concluding remarks are provided in Sect 5.6.

5.2 Related Work

5.2.1 *End, Edge, and Cloud Orchestration for DNN Inference*

Cloud computing significantly promotes the development of AI due to the powerful computing capabilities can be provided for DNN training [34]. Recently, to facilitate efficient DNN inference, the orchestration among end, edge, and cloud has attracted increasing attentions. In [30], an end-cloud synergy DNN inference method is presented for inference accelerating, where the hidden layers of DNN models are respectively processed in the mobile devices or cloud servers such that only intermediate results need to be transmitted. Moreover, Huang et al. [31] proposed a layer-level DNN partitioning strategy and deployed different hidden layers of DNN models among the end, edge, and cloud to distribute the computation loads. In addition to partitioning DNN models, in [32], an edge-cloud orchestration scheme is proposed, in which the shallow hidden layers and deep hidden layers of DNN models are sequentially trained in the cloud and edge servers, thereby the online inference tasks can be processed by the edge. To provide low-latency inference services for intelligent IIoT applications, Yang et al. [33] proposed an end-edge collaborative framework, in which a light-weight and a heavy-weight DNN models are respectively deployed at IIoT devices and MEC servers, such that the computing resources can be efficiently utilized to perform DNN inference.

In short, the reviewed works provide valuable insights to support efficient DNN inference. Different from the above schemes, our work focuses on joint multi-dimensional resource management in end-edge-cloud orchestrated IIoT networks, thereby providing low-latency inference services for intelligent IIoT applications. In our preliminary work [29], we consider a homogeneous industrial scenario, in which a pre-trained DNN model can provide inference services for all intelligent IIoT applications. Hence, spectrum and computing resources are allocated to support the inference task assignment. As an extension of [29], in this work, we consider a heterogeneous industrial scenario where the pre-trained DNN models with differentiated structures are required to support intelligent IIoT applications. In particular, the pre-trained DNN models should be appropriately placed among network nodes to satisfy the diversified inference service requests. Thus, except for spectrum and computing resources, we further consider the caching resource to support the flexible DNN model placement.

5.2.2 *DRL for Resource Scheduling*

DRL algorithms have been extensively applied to address the resource management problems due to the efficient decision-making capability. For instance, Liu et al. [38] proposed a DQN-based algorithm to maximize the long-term computation resource utility of MEC networks. Zhao et al. [39] utilized dueling double DQN to enable efficient user association in cellular networks. In [40], an asynchronous advantage actor critic based scheme has been proposed to achieve adaptive bandwidth allocation in radio access networks, thereby providing reliable transmission for video data streams. In [41], Zhang et al., exploited the deep weighted double Q-learning algorithm to learn a dynamic caching policy in ultra dense networks, with the objective of delay minimization. Moreover, in [35], the algorithm has been leveraged to solve the joint resource optimization problem in vehicular networks.

To sum up, the above DRL-based schemes can vastly improve the resource utilization in wireless networks. Most of which focused on optimizing different network performance metrics, such as communication delay and network throughput. However, our work considers the DNN inference services, in which inference accuracy is a key performance metric, and we focus on leveraging DRL methods to achieve efficient resource management and provide high-accuracy inference services for intelligent IIoT applications.

5.3 System Model and Problem Formulation

In this section, an end-edge-cloud orchestration architecture is introduced to support flexible inference task assignment and DNN model placement, and followed by the communication, caching, and computing models for resource allocation in the considered IIoT networks. Then, a joint optimization problem is formulated under multi-dimensional resource constraints.

5.3.1 End-Edge-Cloud Orchestration Framework

We consider an AI-empowered smart factory, where numerous APLs and industrial facilities are operating, and DNN models are deployed to achieve intelligent IIoT applications, e.g., fault identification. To complete a DNN inference at one node, the node should possess the inference task and its dedicated pre-trained DNN model simultaneously. As shown in Fig. 5.2, we introduce an end-edge-cloud orchestration architecture composed of three layers with different functionalities and resources, i.e., end, edge, and cloud layers.

- In the cloud layer, sufficient computing and caching resources are deployed, thus DNN training is carried out in this layer, and the well-trained DNN models are stored in the cloud caching server (CCaS) as placeable intelligent services.
- In the edge layer, a pair of edge computing/caching server (ECoS/ECaS) are deployed on one BS [48], which can support the inference task execution and DNN model placement, respectively. In addition, an edge controller is installed at

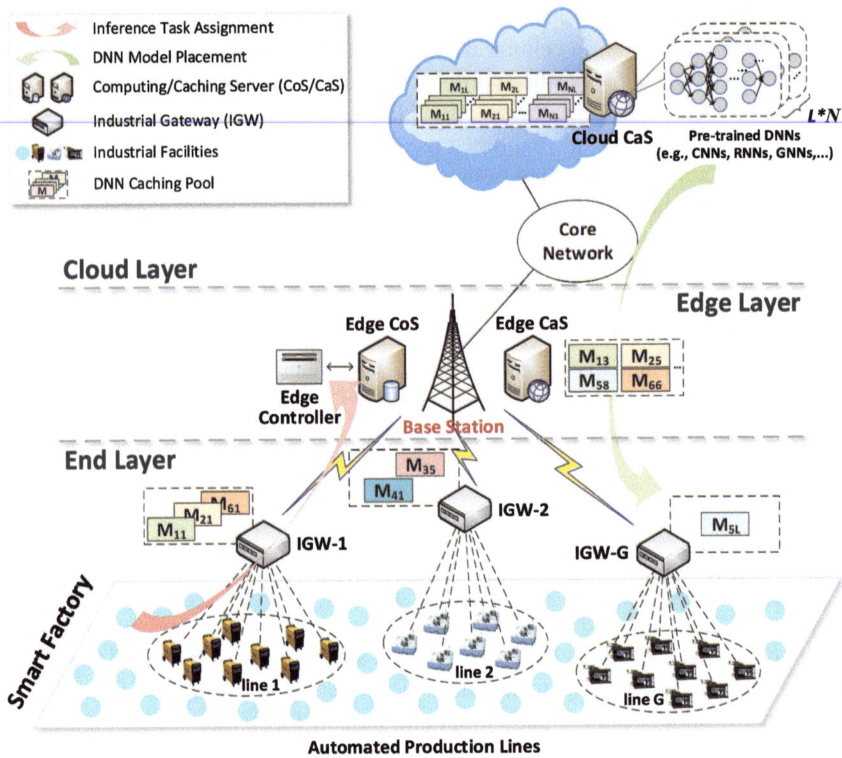

Fig. 5.2 The end-edge-cloud orchestration architecture in IIoT networks

the ECoS which coordinates net-work nodes and observes network status in real time [49].

- In the end layer, IIoT devices collect real-time data from the APLs, and the industrial gateway (IGW) connects the massive inference tasks into the Internet. In each time slot, the edge controller makes decisions to assign inference tasks to the IGWs or ECoS, and place DNN models from the CCaS to the IGWs or ECaS.

We consider a set $\mathcal{K}_i$ of IIoT devices and a set $\mathcal{G}$ of IGWs which covers all IIoT devices, where $\mathcal{K}_i = \{1, 2, \cdots, |\mathcal{K}_i|\}$ and $i \in \mathcal{G} = \{1, 2, \cdots, |\mathcal{G}|\}$. Hence, we define a DNN inference $I_{i,k} \triangleq [W_{i,k}, M^n_{i,k}], \forall i \in \mathcal{G}, k \in \mathcal{K}_i, n \in \{0, 1\}$, where $W_{i,k}$ and $M^n_{i,k}$ respectively denote the generated inference task and the required DNN model, n denotes the accuracy level of DNN models, i.e., a light-weight DNN with low accuracy and a heavy-weight DNN with high accuracy. The details of the DNN inference definition are described in the following parts. A summary of important notations is given in Table 5.1.

Inference Task: In the considered IIoT scenario, massive inference tasks are generated by various applications. Due to the complex industrial environment, the on-site

Table 5.1 Summary of notations

Notation	Description
$D_{i,k}$	Task data size of IIoT device k under IGW i
$T_{i,k}^{\max}$	Maximum tolerance of task completion delay
$\theta_{i,k}$	Data quality of inference task
$P_{i,k}^{n}$	Inference accuracy of each IIoT device with level n
$Q_{i,k}^{n}$	Required workload
$S_{i,k}^{n}$	Model size
R^c, R^e	Transmission rates from cloud to BS and from BS to IGWs
B, B^e	Available bandwidths from device to IGW and from IGW to BS
$\mathcal{C}^l$, $\mathcal{C}^e$	Computing capabilities of IGWs and ECoS
$\mathcal{S}^l$, $\mathcal{S}^e$	Caching capacities of IGWs and ECoS
$u_{i,k}^{l_n}$, $u_{i,k}^{e_n}$	Denote whether the required DNN exists in the IGWs or ECaS
$m_{i,k}$	The binary decision variable of task assignment
$\varphi_{i,k}$	The binary decision variable of DNN model selection
$\alpha_{i,k}^{l}$, $\alpha_{i,k}^{e}$	The decision variables of spectrum allocation
$\beta_{i,k}^{l}$, $\beta_{i,k}^{e}$	The decision variables of computing resource allocation
$\omega_{i,k}^{l_n}$, $\omega_{i,k}^{e_n}$	The decision variables of DNN model placement

data is susceptible to be interfered by the high-frequency noises. Hence, for IIoT device k, an inference task is defined as $W_{i,k} \triangleq (D_{i,k}, T_{i,k}^{\max}, \theta_{i,k}), \forall i \in \mathcal{G}, k \in \mathcal{K}_i$, where $D_{i,k}$ (in bits) denotes the data size, $T_{i,k}^{\max}$ (in seconds) denotes the maximum tolerance of task completion delay, and $\theta_{i,k} \in \{0, 1\}$ denotes the data quality. After the inference task is generated, packet detection is performed first. Here, if the data is noiseless, $\theta_{i,k} = 0$, otherwise $\theta_{i,k} = 1$. Note that, the data size of a task is the same regardless of $\theta_{i,k} = 0$ or 1.

Pre-trained DNN Model: We define a pre-trained DNN model $M_{i,k}^{n} \triangleq (P_{i,k}^{n}, Q_{i,k}^{n}, S_{i,k}^{n}), \forall i \in \mathcal{G}, k \in \mathcal{K}_i, n \in \{0, 1\}$, where $P_{i,k}^{n}$, $Q_{i,k}^{n}$, and $S_{i,k}^{n}$ represent the inference accuracy, the required workload, and the model size, respectively. Considering the heterogeneity of industrial facilities and the limited resources of IIoT networks, the system provides each task with two accuracy levels' dedicated DNN models, which can be selected to perform the corresponding inference tasks according to the available resources. Here, $n = 0$ represents a light-weight DNN model which can provide

Table 5.2 The relationship between pre-trained DNNs and data quality

Data QualityDNNs	Light-weight ($n = 0$)	Heavy-weight ($n = 1$)
Noiseless data($\theta_{i,k} = 0$)	$(\mathcal{P}_{\text{UL}}, S^0_{i,k}, Q^0_{i,k})$	$(\mathcal{P}_{\text{UH}}, S^1_{i,k}, Q^1_{i,k})$
Noisy data($\theta_{i,k} = 1$)	$(\mathcal{P}_{\text{DL}}, S^0_{i,k}, Q^0_{i,k})$	$(\mathcal{P}_{\text{DH}}, S^1_{i,k}, Q^1_{i,k})$

low-accuracy inference services, and $n = 1$ represents a heavy-weight DNN model which can provide high-accuracy inference services. Both levels' DNN models have the same neural network structure, such as fully-connected, recurrent, and convolutional structures, yet are with different number of hidden layers. In addition, the number of hidden layers of heavy-weight DNN models is more than light-weight DNN models. Notably, more hidden layers generally lead to lower approximation errors, such that achieving higher inference accuracy. For each inference task, both accuracy levels of DNNs can be selected to complete the DNN inference, yet a more accurate DNN model occupies more computing and caching resources. Table 5.2 shows the relationship among the inference accuracy, the required workload, and the model size. For simplicity, if $\theta_{i,k} = 1$, then $P^0_{i,k} = \mathcal{P}_{\text{DL}}$ and $P^1_{i,k} = \mathcal{P}_{\text{DH}}$; otherwise, $P^0_{i,k} = \mathcal{P}_{\text{UL}}$ and $P^1_{i,k} = \mathcal{P}_{\text{UH}}$. Note that $\mathcal{P}_{\text{UL}} < \mathcal{P}_{\text{UH}}$, $\mathcal{P}_{\text{DL}} < \mathcal{P}_{\text{DH}}$, $\mathcal{P}_{\text{DL}} < \mathcal{P}_{\text{UL}}$, and $\mathcal{P}_{\text{DH}} < \mathcal{P}_{\text{UH}}$. In addition, $S^0_{i,k} < S^1_{i,k}$ and $Q^0_{i,k} < Q^1_{i,k}$.

At the beginning of each time slot, the edge controller makes the following decisions: (1) where the task should be assigned to, namely task assignment decision; (2) which pre-trained DNN model should be selected and whether it is cached at the corresponding nodes, namely DNN model placement decision; and (3) how much spectrum, computing, and caching resources should be allocated, namely resource allocation decisions. In each time slot, we assume that each IIoT device generates one task, so there are $N = \sum_{i=1}^{G} |\mathcal{K}_i|$ tasks need to be processed by their dedicated pre-trained DNN models. Considering the limited resources of IGWs, the tasks have to be collaboratively performed within the end-edge-cloud orchestration architecture, thereby completing inference tasks within the strict delay requirements.

5.3.2 *Communication Model*

Due to the long distance between IIoT devices and BS, massive sensing data of IIoT devices is aggregated by IGWs. The IGWs further access the aggregated data by cellular network techniques that can support reliable long-distance data transmission. For communication between IIoT devices and IGWs, we consider the industrial wireless sensor network technique [42]. For communication between IGWs and BS, we consider orthogonal frequency division multiple access technique to transmit inference tasks from the IGWs to the BS. Both networks are equipped with the available bandwidth B and B^e, respectively. The task $W_{i,k}$ in each time slot can only be assigned to one node, i.e., IGW i or ECoS. Let $m_{i,k} \in \{0, 1\}$ denote the binary decision variable of task assignment. Here, if $m_{i,k}$ is 1, $W_{i,k}$ is accordingly assigned to IGW i, otherwise to ECoS. We assume the communication between each IIoT device and each IGW is assigned an orthogonal channel, i.e., no interference exists in the network. Thus, the spectrum efficiency for IIoT devices is given by

$$\gamma_{i,k} = \log_2\left(1 + \frac{p_{i,k}|h_{i,k}|^2}{\sigma^2}\right), \forall i \in \mathcal{G}, k \in \mathcal{K}_i. \tag{5.1}$$

Similarly, $\gamma^e_{i,k} = \log_2\left(1 + \frac{p^e_{i,k}|h^e_{i,k}|^2}{\sigma^2}\right)$ represents the spectrum efficiency for IGWs. Here, $p_{i,k}$ and $p^e_{i,k}$ denote the transmission powers of IIoT device k and IGW i, $\left|h_{i,k}\right|^2$ and $|h^e_{i,k}|^2$ denote the gains of the IIoT device-IGW and IGW-BS channels, and σ^2 denotes the power of the Gaussian noise. Hence, the transmission rate for device k is calculated by

$$R_{i,k} = \alpha^l_{i,k} B \gamma_{i,k}, \forall i \in \mathcal{G}, k \in \mathcal{K}_i. \tag{5.2}$$

Similarly, $R^e_{i,k} = \alpha^e_{i,k} B^e \gamma^e_{i,k}$ represents the transmission rate for IGW i. IIoT device k and IGW i are allocated a fraction $\alpha^l_{i,k}$ of bandwidth B and a fraction $\alpha^e_{i,k}$ of bandwidth B^e, respectively, where $0 \le \alpha^l_{i,k}, \alpha^e_{i,k} \le 1$, $\sum_{i\in\mathcal{G},k\in\mathcal{K}_i} m_{i,k}\alpha^l_{i,k} \le 1$, and $\sum_{i\in\mathcal{G},k\in\mathcal{K}_i} (1 - m_{i,k})\alpha^e_{i,k} \le 1$. Thus, the transmission delay for offloading an inference task from IIoT device k to the connected IGW i and from device k to the BS can be calculated by

$$t^l_{i,k} = \frac{D_{i,k}}{R_{i,k}}, \forall i \in \mathcal{G}, k \in \mathcal{K}_i, \tag{5.3}$$

and

$$t^e_{i,k} = \frac{D_{i,k}}{R_{i,k}} + \frac{D_{i,k}}{R^e_{i,k}}, \forall i \in \mathcal{G}, k \in \mathcal{K}_i, \tag{5.4}$$

respectively. Here, we ignore the transmission delay on wired communications between the BS and ECoS due to the wired connection. In addition, the size of the inference result is much smaller than that of the task [33], hence, the time consumption on returning the inference result to the APLs can be neglected in this chapter.

5.3.3 Caching Model

As depicted in Section III-A, all the pre-trained DNN models are stored in the CCaS. In addition, the IGWs and ECaS are respectively equipped with $\mathcal{S}^l$ and $\mathcal{S}^e$ caching spaces for caching DNN models. During the system operation, the pre-trained DNN models are placed to the IGWs or ECaS from the CCaS according to the requests. For each inference task, a heavy-weight and a light-weight DNN models can be provided for inference services. In each time slot, the system selects an appropriate DNN model for each inference task. Hence, we define $\varphi_{i,k} \in \{0, 1\}, \forall i \in \mathcal{G}, k \in \mathcal{K}_i$, to represent which level of DNN model is selected, where $\varphi_{i,k} = 1$ indicates a heavy-weight DNN and $\varphi_{i,k} = 0$ indicates a light-weight DNN. Note that heavy-weight DNN models have higher inference accuracy, yet occupying more caching resources.

Let binary variables $u_{i,k}^{l_n}, u_{i,k}^{e_n} \in \{0, 1\}, \forall i \in \mathcal{G}, k \in \mathcal{K}_i, n \in \{0, 1\}$ denote whether the required DNN model exists in the IGWs or ECaS. Taking $u_{i,k}^{l_0}$ as an example, $u_{i,k}^{l_0} = 0$ indicates that the required light-weight DNN does not exist in the corresponding IGWs, otherwise is the opposite. If an inference task is assigned to a node and an appropriate pre-trained DNN model is selected, the edge controller then checks the corresponding caching pools to make sure whether the required DNN has been stored. If $u_{i,k}^{l_n}$ or $u_{i,k}^{e_n}$ is 0, a DNN model placement request of a node in low layers is sent to the superior nodes, and then the required DNN models can be transmitted to the node that sent the request. Note that the request is unidirectional, namely, the required DNN models can only be transmitted from CCaS to ECaS, from CCaS to IGW, or from ECaS to IGW. Hence, the transmission delay of $M_{i,k}^n$ from CCaS to ECaS is given by

$$\tau_{i,k}^{c_n} = \frac{S_{i,k}^n}{R^c}, \forall i \in \mathcal{G}, k \in \mathcal{K}_i. \tag{5.5}$$

Similarly, $\tau_{i,k}^{e_n} = \frac{S_{i,k}^n}{R^e}$. Thus, $\tau_{i,k}^{l_n} = \tau_{i,k}^{c_n} + \tau_{i,k}^{e_n}$, where $\tau_{i,k}^{e_n}$ and $\tau_{i,k}^{l_n}$ represent the transmission delays from ECaS to IGW, and from CCaS to IGW, respectively. In addition, R^c and R^e denote the transmission rates from cloud to the BS and from BS to the IGWs, respectively.

Let binary decision variables $\omega_{i,k}^{l_n}, \omega_{i,k}^{e_n} \in \{0, 1\}, \forall i \in \mathcal{G}, k \in \mathcal{K}_i$ denote whether cache the requested DNN models in the corresponding caching pools after finishing the inference tasks, where $n = 0$ or 1 represents the caching decision for a light-weight or heave-weight DNN model. Taking $\omega_{i,k}^{l_n}$ as an example, $\omega_{i,k}^{l_n} = 1$ indicates that the DNN models can be cached in the corresponding IGWs to provide instant inference services for the following tasks, otherwise is the opposite. With these caching decisions, the pre-trained DNN models can be flexibly placed at the end device and edge server, and the caching resources of network nodes can be utilized more reasonably.

5.3.4 Computing Model

The IGWs and ECoS are equipped with the computing capabilities of C^l and C^e (CPU cycles per second), respectively. As depicted in Table 5.2, a heavy-weight DNN model achieves high inference accuracy while consuming more computing resource, and a light-weight DNN is the opposite. Let $\beta_{i,k}^l$, and $\beta_{i,k}^e (0 \leq \beta_{i,k}^l, \beta_{i,k}^e \leq 1)$ denote the fractions of computing resources allocated to the inference tasks, respectively. Thus, the delay for inference task executing at IGWs can be calculated by

$$c^l_{i,k} = \frac{Q^n_{i,k}}{\beta^l_{i,k} C^l}, \forall i \in \mathcal{G}, k \in \mathcal{K}_i. \tag{5.6}$$

Similarly, $c^e_{i,k} = \frac{Q^n_{i,k}}{\beta^e_{i,k} C^e}$ for task executing at ECoS. In addition, the decision variables of computing resources are constrained by $\sum_{i \in \mathcal{G}, k \in \mathcal{K}_i} m_{i,k} \beta^l_{i,k} \le 1$ and $\sum_{i \in \mathcal{G}, k \in \mathcal{K}_i} (1 - m_{i,k}) \beta^e_{i,k} \le 1$.

Accordingly, the inference task completion delay at IGWs and ECoS can be calculated by

$$\begin{aligned} d^l_{i,k} = &\max\{t^l_{i,k}, (1 - u^{l_n}_{i,k}) u^{e_n}_{i,k} \tau^{e_n}_{i,k} + (1 - u^{l_n}_{i,k}) \cdot \\ &* (1 - u^{e_n}_{i,k}) \tau^{l_n}_{i,k}\} + c^l_{i,k}, \forall i \in \mathcal{G}, k \in \mathcal{K}_i, \end{aligned} \tag{5.7}$$

and

$$d^e_{i,k} = \max\{t^e_{i,k}, (1 - u^{e_n}_{i,k}) \tau^{c_n}_{i,k}\} + c^e_{i,k}, \forall i \in \mathcal{G}, k \in \mathcal{K}_i, \tag{5.8}$$

respectively. Here, for $d^l_{i,k}$ or $d^e_{i,k}$, due to the inference task and the required DNN model are simultaneously transmitted to a node, thus the maximum value between the transmission delays of inference task and DNN model is considered as the corresponding total transmission delay.

Taking two cases into consideration, the total time for completing the inference task $W_{i,k}$ can be calculated by

$$d_{i,k} = m^l_{i,k} d^l_{i,k} + m^e_{i,k} d^e_{i,k}, \forall i \in \mathcal{G}, k \in \mathcal{K}_i. \tag{5.9}$$

Once the DNN selecting decision $\varphi_{i,k}$ is determined and the data quality $\theta_{i,k}$ is detected, the task can be executed and then each IIoT device obtains an inference accuracy, i.e.,

$$\begin{aligned} U_{i,k} = &\left(1 - \theta_{i,k}\right) [\varphi_{i,k} \mathcal{P}_{\mathrm{UH}} + \left(1 - \varphi_{i,k}\right) \mathcal{P}_{\mathrm{UL}}] + \\ &\theta_{i,k} [\varphi_{i,k} \mathcal{P}_{\mathrm{DH}} + \left(1 - \varphi_{i,k}\right) \mathcal{P}_{\mathrm{DL}}], \forall i \in \mathcal{G}, k \in \mathcal{K}_i. \end{aligned} \tag{5.10}$$

Note that the corresponding accuracy can be obtained only when the task is completed within the delay requirement.

5.3.5 Optimization Problem Formulation

In the considered IIoT networks, we deploy an end-edge-cloud orchestration architecture to support DNN inference, in which the inference tasks of IIoT devices are assigned to the IGWs and ECoS, while the pre-trained DNN models are placed to the IGWs and ECaS from CCaS. In this chapter, we focus on enabling high-accuracy

DNN inference in IIoT networks. As previously described, inference accuracy is a key performance metric for DNN-based IIoT applications. Only the task is completed within strict delay requirements, the system then can obtain the corresponding accuracy. Enabling the tasks to be reasonably assigned among network nodes to obtain more inference accuracies is important to improve the overall system performance. Therefore, aiming at maximizing the average accuracy of inference tasks, while satisfying the delay requirements of tasks and the constraints of spectrum, caching, and computing resources, we formulate the optimization problem as follows:

$$\max_{\boldsymbol{m},\boldsymbol{\varphi},\boldsymbol{\alpha},\boldsymbol{\beta},\boldsymbol{\omega}} \frac{1}{N}\sum_{i=1}^{G}\sum_{k=1}^{|\mathcal{K}_i|} U_{i,k} \tag{5.11}$$

and subject to:

$$\begin{aligned}
&C1: d_{i,k} \le T_{i,k}^{\max}, \forall i \in \mathcal{G}, k \in \mathcal{K}_i,\\
&C2: m_{i,k} \in \{0,1\}, \forall i \in \mathcal{G}, k \in \mathcal{K}_i,\\
&C3: \varphi_{i,k} \in \{0,1\}, \forall i \in \mathcal{G}, k \in \mathcal{K}_i,\\
&C4: \sum_{i\in\mathcal{G},k\in\mathcal{K}_i} m_{i,k}\alpha_{i,k}^{l} \le 1, \sum_{i\in\mathcal{G},k\in\mathcal{K}_i} (1-m_{i,k})\alpha_{i,k}^{e} \le 1,\\
&\quad\;\; 0 \le \alpha_{i,k}^{l}, \alpha_{i,k}^{e} \le 1,\\
&C5: \sum_{i\in\mathcal{G},k\in\mathcal{K}_i} m_{i,k}\beta_{i,k}^{l} \le 1, \sum_{i\in\mathcal{G},k\in\mathcal{K}_i} (1-m_{i,k})\beta_{i,k}^{e} \le 1,\\
&\quad\;\; 0 \le \beta_{i,k}^{l}, \beta_{i,k}^{e} \le 1,\\
&C6: \sum_{i\in\mathcal{G}}\sum_{k\in\mathcal{K}_i} m_{i,k}[\varphi_{i,k}S_{i,k}^{1} + (1-\varphi_{i,k})\, S_{i,k}^{0}] \le \mathcal{S}^{l},\\
&\quad\;\; \sum_{i\in\mathcal{G}}\sum_{k\in\mathcal{K}_i} (1-m_{i,k})[\varphi_{i,k}S_{i,k}^{1} + (1-\varphi_{i,k})\, S_{i,k}^{0}] \le \mathcal{S}^{e},\\
&C7: \omega_{i,k}^{l_0}, \omega_{i,k}^{l_1}, \omega_{i,k}^{e_0}, \omega_{i,k}^{e_1} \in \{0,1\}, \forall i \in \mathcal{G}, k \in \mathcal{K}_i,
\end{aligned}$$

where $U_{i,k}$ is given by Eq. (5.10) and N is the number of inference tasks in each time slot. In addition, the optimization variables consist of five components: (1) $\boldsymbol{m} = \{m_{i,k}\}$ is the task assignment decision matrix for assigning inference tasks to IGWs and ECoS; (2) $\boldsymbol{\varphi} = \{\varphi_{i,k}\}$ is the matrix describing the multi-level DNN selection decisions; (3) $\boldsymbol{\alpha} = \{\alpha_{i,k}^{l}, \alpha_{i,k}^{e}\}$ is the matrix describing spectrum resources allocated to inference tasks by the IGWs and BS; (4) $\boldsymbol{\beta} = \{\beta_{i,k}^{l}, \beta_{i,k}^{e}\}$ is the matrix describing computing resources allocated to inference tasks by the IGWs and ECoS; and (5) $\boldsymbol{\omega} = \{\omega_{i,k}^{l_0}, \omega_{i,k}^{l_1}, \omega_{i,k}^{e_0}, \omega_{i,k}^{e_1}\}$ is the matrix describing whether cache the requested DNN models in the corresponding nodes. For constraints, $C1$ first guarantees that the entire time cost of DNN inference cannot exceed the maximum delay tolerance $T_{i,k}^{\max}$ of $W_{i,k}$. $C2$ and $C3$ ensure that the inference task can only be executed at one node and only one pre-trained DNN model can be selected in each time slot.

Moreover, $C4 - C6$ indicate that the resource allocation decisions should be made with the available bandwidth (B and B^e), computing ($\mathcal{C}^l$ and $\mathcal{C}^e$), and caching ($\mathcal{S}^l$ and $\mathcal{S}^e$) resources, respectively. $C7$ indicates whether the requested DNN models can be stored in the corresponding nodes or not. By solving the formulated problem, the pre-trained DNNs can be flexibly placed and selected based on the delay requirements and available multi-dimensional resources, thereby effectively supporting DNN inference.

5.4 DRL-Based Resource Scheduling Scheme for DNN Inference

According to the Eq. (5.11), the formulated TADP problem is a mix-integer programming problem, which is difficult to be solved by traditional optimization methods. Moreover, due to the strict delay requirements of inference tasks, the problem should be solved quickly. To this end, we propose a DRL-based scheme to achieve efficient resource management, thereby obtaining the solution rapidly. In this section, we first transform the formulated problem into an MDP, and then propose a learning-based algorithm to solve the formulated optimization problem efficiently.

5.4.1 MDP Transformation

Considering the mix-integer variables of the formulated optimization problem and the strict delay requirements of inference tasks, instead of using traditional optimization methods, we present a learning-based algorithm for efficient resource management to achieve DNN inference in IIoT networks [36]. To rapidly obtain the optimal decisions under stochastic environment, we first transform the formulated problem into an MDP, and then adopt a DRL-based approach to solve it [37]. The framework of DRL is composed of the agent and environment. As shown in Fig. 5.2, the agent is coordinated by the edge controller installed at the ECoS, and everything beyond that is deemed as the environment. The key components of the DRL method, i.e., the observed state, action, and reward, are summarized as follows:

1. *State*: At each time slot, IIoT devices and network nodes periodically send the task information $\{D_{i,k}, T_{i,k}^{\max}, \theta_{i,k}\}$ and caching pool state information $\{u_{i,k}^{l_0}, u_{i,k}^{l_1}, u_{i,k}^{e_0}, u_{i,k}^{e_1}\}$ to the ECoS. Thus, the agent can obtain the state space $\mathcal{S}$ of the considered IIoT system. Then, the state vector at time slot t, $s(t) \in \mathcal{S}$, can be defined as

$$
\begin{aligned}
s(t) = \{&\boldsymbol{D}_{i,k}(t), \boldsymbol{T}_{i,k}^{\max}(t), \boldsymbol{\theta}_{i,k}(t), \boldsymbol{u}_{i,k}^{l_0}(t), \boldsymbol{u}_{i,k}^{l_1}(t), \\
&\boldsymbol{u}_{i,k}^{e_0}(t), \boldsymbol{u}_{i,k}^{e_1}(t)\}, \forall i \in \mathcal{G}, k \in \mathcal{K}_i.
\end{aligned}
\tag{5.12}
$$

2. *Action*: According to the observed environment states, the agent makes decisions based on the policy π. The decisions, including the task assignment $m_{i,k}$, multi-levels DNN selection $\varphi_{i,k}$, resource allocation ratios of spectrum $\{\alpha_{i,k}^{l}, \alpha_{i,k}^{e}\}$ and computing $\{\beta_{i,k}^{l}, \beta_{i,k}^{e}\}$, and DNN model placement $\{\omega_{i,k}^{l_0}, \omega_{i,k}^{l_1}, \omega_{i,k}^{e_0}, \omega_{i,k}^{e_1}\}$, made by the agent at time slot t, $a(t) \in \mathcal{A}$, can be described as

$$\begin{aligned} a(t) = \{&m_{i,k}(t), \boldsymbol{\varphi}_{i,k}(t), \boldsymbol{\alpha}_{i,k}^{l}(t), \boldsymbol{\alpha}_{i,k}^{e}(t), \boldsymbol{\beta}_{i,k}^{l}(t), \boldsymbol{\beta}_{i,k}^{e}(t), \\ &\boldsymbol{\omega}_{i,k}^{l_0}(t), \boldsymbol{\omega}_{i,k}^{l_1}(t), \boldsymbol{\omega}_{i,k}^{e_0}(t), \boldsymbol{\omega}_{i,k}^{e_1}(t)\}, \forall i \in \mathcal{G}, k \in \mathcal{K}_i, \end{aligned} \tag{5.13}$$

where $a(t) \in \mathcal{A}$ is with $N \times 10$ dimensions.

3. *Reward*: Once action $a(t)$ is taken by the environment based on the observed state $s(t)$, the agent then will obtain a reward. In the training stage, the policy π is iterated according to the obtained reward until achieving the convergence. To maximize the accuracy of DNN inference, we therefore define the reward for inference tasks at time slot t, i.e.,

$$r(t) = \begin{cases} \frac{1}{N} \sum_{i=1}^{G} \sum_{k=1}^{|\mathcal{K}_i|} U_{i,k}(t), & \text{if } C1 - C7 \text{ are satisfied,} \\ 0, & \text{otherwise,} \end{cases} \tag{5.14}$$

where $U_{i,k}(t)$ is the accuracy of inference task $W_{i,k}$ in time slot t. The reward function indicates that only the appropriate resources are allocated and the delay requirement are satisfied, simultaneously, the agent then can obtain the system reward. Otherwise, the task cannot be completed, thereby $r(t) = 0$. For notational simplicity, the time index t is abbreviated as the subscript in the following context.

5.4.2 TD3-Based Resource Scheduling Algorithm

In the considered IIoT networks, it is difficult to model the environment state accurately due to the massive inference requests and the unknown caching pool state transitions. Moreover, the high-dimensional state and action spaces caused by the network scale (i.e., G and N) and multi-dimensional resources, also result in intractable convergence performance [35]. In recent years, actor-critic methods, such as DDPG [43], have obtained considerable achievements in solving optimization problems with large action spaces or continuous decision variables [24]. TD3, as an enhanced version of DDPG, effectively addresses the function approximation error to reduce overestimation bias and avoid suboptimal policy updates, thereby improving the learning speed and performance in continuous control domains [44]. In this section, we propose a TD3-based algorithm to solve the transformed MDP problem.

For the proposed TD3-based algorithm, the goal of the agent is to find an optimal policy π^* with the maximum average inference accuracy. During the learning stage,

Algorithm 1: TD3-Based Algorithm for Multi-Dimensional Resource Management

Initialize actor network $\pi_{\boldsymbol{\theta}}(s)$ and critic networks Q_{ω_1}, Q_{ω_2} with parameters $\boldsymbol{\theta}, \boldsymbol{\omega}_1, \boldsymbol{\omega}_2$;
Initialize target networks with weights $\boldsymbol{\theta}', \boldsymbol{\omega}_1', \boldsymbol{\omega}_2'$;
Initialize experience replay buffer $\mathcal{D}_r$;
for *episode* $= 1$ *to* E **do**
  Initialize environment, and receive initial state s_1;
  for *step* $t = 1$ *to* T **do**
    Select action with the current policy $a_t \sim \pi_{\boldsymbol{\theta}}(s_t) + \epsilon, \epsilon \sim \mathcal{N}(0, \rho)$;
    ▷ DNN caching pool updating
    Discretize action $m_{i,k}, \varphi_{i,k}, \omega_{i,k}^{l_0}, \omega_{i,k}^{l_1}, \omega_{i,k}^{e_0}, \omega_{i,k}^{e_1}$;
    for *IGWs* $i = 1$ *to* G **do**
      for *IIoT devices* $k = 1$ *to* $|\mathcal{K}_i|$ **do**
        • Step 1: Compute remaining caching space
        if $u_{i,k}^{l_0} = 1$ **then**
          $S^l[i] = S^l[i] + S_{i,k}^n$
        • Step 2: Compute transmission delay $\tau_{i,k}$
        if $m_{i,k} = 1, \varphi_{i,k} = 0, u_{i,k}^{l_0} = 0, S^l[i] + S_{i,k}^0 \le \mathcal{S}^l$, *and* $\tau_{i,k} \le T_{i,k}^{\max}$ **then**
          $u_{i,k}^{l_0} = 1, S^l[i] = S^l[i] + S_{i,k}^0$
        else
          $u_{i,k}^{l_0} = 0$
        Update $u_{i,k}^{l_1}, u_{i,k}^{e_0}, u_{i,k}^{e_1}$;
        Execute the inference task;
        • Step 3: Compute completion delay $d_{i,k}$
        if $m_{i,k} = 1, \varphi_{i,k} = 0, u_{i,k}^{l_0} = 0, S^l[i] + S_{i,k}^0 \le \mathcal{S}^l$, *and* $d_{i,k} \le T_{i,k}^{\max}$ **then**
          $u_{i,k}^{l_0} = 1, S^l[i] = S^l[i] + S_{i,k}^0$
        else
          $u_{i,k}^{l_0} = 0$
        Update $u_{i,k}^{l_1}, u_{i,k}^{e_0}, u_{i,k}^{e_1}$;
        • Step 4: Perform caching actions $\{\omega_{i,k}^{l_n}, \omega_{i,k}^{e_n}\}$
        if $u_{i,k}^{l_0} = 1$ *and* $\omega_{i,k}^{l_0} = 1$ **then**
          $u_{i,k}^{l_0} = 1$
        else
          $u_{i,k}^{l_0} = 0$
        Update $u_{i,k}^{l_1}, u_{i,k}^{e_0}, u_{i,k}^{e_1}$;
    ▷ Learning algorithm iterating
    Obtain the reward r_t and observe next state s_{t+1};
    Store the experience $\{s_t, a_t, r_t, s_{t+1}\}$ into the $\mathcal{D}_r$, and replace the oldest ones if the buffer is full;
    Randomly sample a minibatch of D_b experiences $\{s_q, a_q, r_q, s_{q+1}\}$ from $\mathcal{D}_r$;
    Set $y_q = r_q + \gamma \min_{j=1,2} Q_{\omega_j'}(s_{q+1}, \pi_{\theta'}(s_{q+1}) + \epsilon), \epsilon \sim \text{clip}(\mathcal{N}(0, \rho), -c, c)$;
    Update $\boldsymbol{\omega}_1, \boldsymbol{\omega}_2$ by minimizing the loss in Eq. (5.16);
    if t *mod* d **then**
      Update $\boldsymbol{\theta}$ via the policy gradient in Eq. (5.19);
      Update target networks by Eq. (5.21).

the agent randomly samples a minibatch experiences from $\mathcal{D}_r$. Taking the qth experience $\{s_q, a_q, r_q, s_{q+1}\}$ as an example, the actor makes TADP actions $a = \pi_{\theta}(s_q)$ and $a' = \pi_{\theta'}(s_{q+1})$ via the evaluation and target networks with parameters $\boldsymbol{\theta}$ and $\boldsymbol{\theta}'$. Then, the evaluation critics evaluate a and a_q with parameters $\boldsymbol{\omega}_j$, $j \in \{1, 2\}$, and the target critics evaluate a' with parameter $\boldsymbol{\omega}'_j$, to calculate the $Q(s_q, a_q; \boldsymbol{\omega}_j)$ and $Q(s_{q+1}, \pi_{\theta'}(s_{q+1}); \boldsymbol{\omega}'_j)$, respectively. Thus, the target Q-values are given by

$$y_q = r_q + \gamma \min_{j=1,2} Q(s_{q+1}, \pi_{\theta'}(s_{q+1}) + \epsilon; \boldsymbol{\omega}'_j), \tag{5.15}$$

where $\epsilon \sim clip(\mathcal{N}(0, \rho), -c, c)$ denotes the added noise, c denotes the clip bound of ϵ, and $\gamma \in [0, 1]$ denotes a discounting factor. Thus, the loss function of the evaluation critics can be expressed by

$$L(\boldsymbol{\omega}_j) = \frac{1}{2D_b} \sum_{q \in \mathcal{D}_b} \left(y_q - \min_{j=1,2} Q(s_q, a_q; \boldsymbol{\omega}_j) \right)^2, \tag{5.16}$$

where D_b is the size of the sampled minibatch experiences from D_r. Then, parameter $\boldsymbol{\omega}_j$ can be updated by minimizing the loss function, i.e.,

$$\nabla L(\boldsymbol{\omega}_j) = -\frac{1}{D_b} \sum_{q \in \mathcal{D}_b} \left(y_q - \min_{j=1,2} Q(s_q, a_q; \boldsymbol{\omega}_j) \right) \cdot \nabla_{\boldsymbol{\omega}_j} Q(s_q, a_q; \boldsymbol{\omega}_j). \tag{5.17}$$

Moreover, the objective function of the evaluation actor can be expressed by

$$J(\boldsymbol{\theta}) = \frac{1}{D_b} \sum_{q \in \mathcal{D}_b} Q(s_q, a_q; \boldsymbol{\omega}_1). \tag{5.18}$$

Although two critics are deployed in the proposed scheme, only one critic, either Critic1 or Critic2, needs to be used to perform policy update. Here, the policy update depends on the value estimations of Critic1. Thus, the parameter $\boldsymbol{\theta}$ of the evaluation actor can be updated via policy gradient theorem, i.e.,

$$\nabla J(\boldsymbol{\theta}) = \frac{1}{D_b} \sum_{q \in \mathcal{D}_b} \nabla_{\pi(s_q)} Q(s_q, \pi(s_q); \boldsymbol{\omega}_1) \nabla_{\boldsymbol{\theta}} \pi(s_q). \tag{5.19}$$

Hence, the parameters $\boldsymbol{\omega}_j$ and $\boldsymbol{\theta}$ of the evaluation critics and actor can be updated by the gradient descent method, i.e.,

$$\begin{aligned} \boldsymbol{\omega}_j &= \boldsymbol{\omega}_j - \epsilon_c \nabla L(\boldsymbol{\omega}_j), \\ \boldsymbol{\theta} &= \boldsymbol{\theta} - \epsilon_a \nabla J(\boldsymbol{\theta}), \end{aligned} \tag{5.20}$$

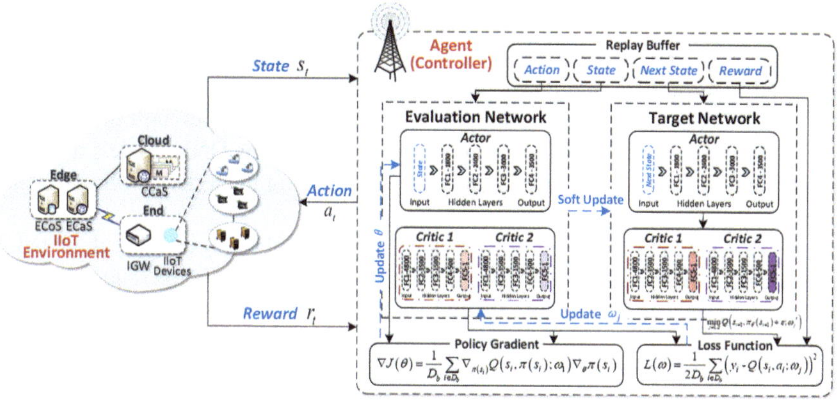

Fig. 5.3 The framework of the proposed TD3-based algorithm

where ϵ_a and ϵ_c denote the learning rate of the actor and critics, respectively. In addition, the soft-updating technique is used for the parameter updating of target networks, i.e.,

$$\begin{aligned} \boldsymbol{\omega}'_j &= \tau\boldsymbol{\omega}_j + (1-\tau)\boldsymbol{\omega}'_j, \\ \boldsymbol{\theta}' &= \tau\boldsymbol{\theta} + (1-\tau)\boldsymbol{\theta}', \end{aligned} \tag{5.21}$$

where τ denotes the update factor. Notably, the updates of policy and target networks are only occurred after d times updates to the evaluation critics, and thus effectively limits the likelihood of repeating updates with an unchanged critic. The proposed TD3-based algorithm for solving the formulated TADP problem in real time can be summarized in Algorithm 1.

Main Techniques of the Proposed TD3-Based Solution: As described in Algorithm 1, a DNN caching pool updating (DCPU) subroutine is proposed to achieve efficient caching resource utilization. In each time slot, once the DNN models requested by inference tasks are sent to the nodes, a certain caching resource needs to be allocated to cache the DNN models in the corresponding caching pools. To improve the flexibility of the system for DNN model placement, we propose a DCPU subroutine to support flexible DNN model placement among end, edge, and cloud, thereby ensure DNN inference can be completed within strict delay requirements. The designed DCPU subroutine consists of the following steps.

Step 1 (Lines 12–14): At the beginning of each time slot, the edge controller first checks the state of DNN caching pools, and allocates caching resources for the DNN models cached in the previous time slot. Thus, the remaining caching spaces of network nodes can be calculated and then can be allocated to the requested DNN models that are transmitted from the ECaS or CCaS.

Step 2 (Lines 15–21): According to the amounts of spectrum and computing resources allocated to the inference tasks, the DNN transmission delay $\tau_{i,k}^{c_n}$, $\tau_{i,k}^{e_n}$, and $\tau_{i,k}^{l_n}$ can be calculated based on the sending location of the requested DNN models. If $\tau_{i,k}^{c_n}$, $\tau_{i,k}^{e_n}$, or $\tau_{i,k}^{l_n}$ is smaller than $T_{i,k}^{\max}$, which means the requested DNNs can be placed to the corresponding caching pools within the delay requirement, thus the DNN models can be cached in this time slot. In such cases, although the inference tasks cannot be guaranteed to be completed within $T_{i,k}^{\max}$, the DNN models requested by current tasks can provide instant inference services for the tasks of following time slots.

Step 3 (Lines 22–27): Similarly, the task completion delay $d_{i,k}$ can be calculated. If $d_{i,k}$ is smaller than the delay requirement of inference task $T_{i,k}^{\max}$, the requested DNN model can be cached in the caching pools to be provided for executing the inference tasks.

Step 4 (Lines 28–33): According to the decision variables $\omega_{i,k}^{l_n}$ and $\omega_{i,k}^{e_n}$, the states of caching pools can be updated, where $\omega_{i,k}^{l_n} = 1$ or $\omega_{i,k}^{e_n} = 1$ indicates that the DNN models requested by inference tasks are continuously cached in the IGWs or ECaS, otherwise indicates that the requested DNN models are deleted from the corresponding nodes.

In addition, as shown in Fig. 5.3, the proposed learning algorithm is supported by three main techniques:

1. The experience replay technique is adopted for accelerating the convergence efficiency in an off-policy way, in which the experience of the agent at each time slot $e_t = \{s_t, a_t, r_t, s_{t+1}\}$ is stored in a replay buffer $\mathcal{D}_r = \{e_1, e_2, \ldots, e_r\}$ with a certain capacity. When the training process is started, the first D_r experiences are stored in the replay buffer. During the time slots of the learning stage, the agent randomly samples a minibatch of experiences from the replay buffer, thus the correlation among the samples can be effectively reduced.
2. The "double-network" (i.e., evaluation and target networks) mechanism is utilized to stabilize the learning process, and the actor-critic framework is adopted in each network to handle the high-dimensional and continuous action space. Moreover, the clipped double Q-learning for actor-critic is adopted to reduce the overestimation bias. Hence, the agent is equipped with six function approximators, namely, evaluation actor/Critic1/Critic2 and target actor/Critic1/Critic2. In each time slot, the evaluation actor firstly makes resource allocation decisions according to the observed environment state and the policy π. The evaluation critics (i.e., Critic1 and Critic2) evaluate the decisions by generating a pair of Q-values, and the minimum between the two estimates is selected to update the pair of critics. The target network is responsible for calculating a pair of target Q-values and the minimum between the target Q-values is provided to the evaluation network for parameter iterating, such that mitigating overestimation bias and improve the algorithm performance. And then, the target network periodically updates the latest network parameters from the evaluation network. Notably, we deploy multi-layer FCNNs as the function approximators of the proposed TD3-based learning algorithm.

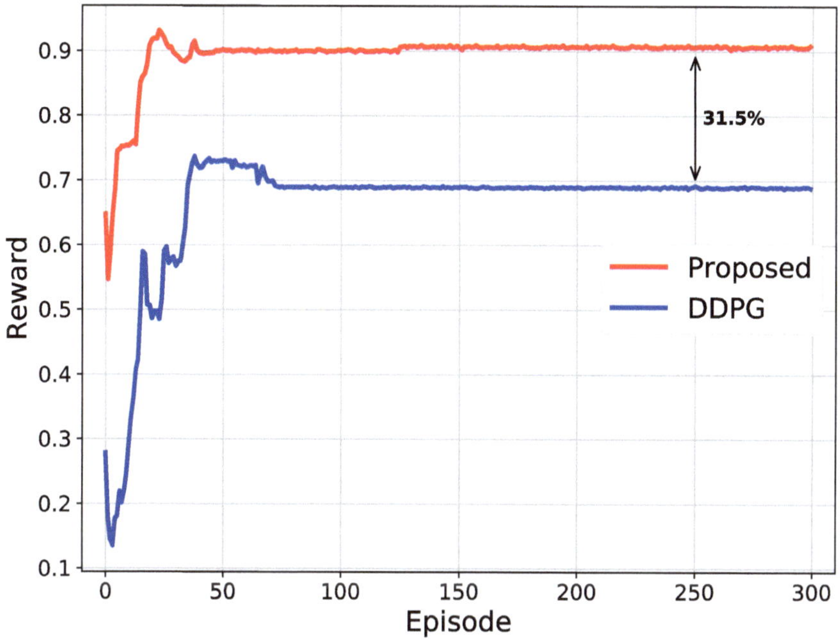

Fig. 5.4 The convergence performance of the proposed algorithm

3. The delayed policy update and target policy smoothing regularization are used to reduce per-update error and avoid overfitting in the value estimation. Generally, deep function approximators require massive gradient iterations to converge. However, policy updates with high-error Q-values often result in divergent behavior, thus the evaluation actor should be updated at a lower frequency than the evaluation critics until the value error is neglectable. In the proposed TD3-based algorithm, the policy and target networks can be updated only after d updates to the evaluation critics. Additionally, when updating the critic networks, the target Q-values are very susceptible to inaccuracies caused by function approximation error, thereby increasing the variance of the target Q-values. Thus, a clipped random noise is added to further improve the accuracy of Q-value estimation and the stability of the learning process.

Remark: The TD3-based algorithm is proposed to achieve efficient resource utilization, which is different from existing learning approaches. The difference is two-fold: (1) The proposed TD3-based resource management scheme is endowed with the DCPU subroutine to achieve more flexible task assignment and DNN placement; and (2) The algorithm is enhanced with three important techniques, i.e., dual critics, delayed policy update and target policy smoothing regularization, which can reduce the overestimation bias and stabilize the learning process effectively, as compared to other learning approaches.

Table 5.3 Parameters of pre-trained DNN models

Parameter	Value	Parameter	Value
$\mathcal{P}_{\mathrm{UL}}$	[0.915, 0.938]	$\mathcal{P}_{\mathrm{UH}}$	[0.973, 0.992]
$\mathcal{P}_{\mathrm{DL}}$	[0.876, 0.897]	$\mathcal{P}_{\mathrm{DH}}$	[0.929, 0.954]
$S_{i,k}^{0}$	[3.5, 5.5] MB	$S_{i,k}^{1}$	[9, 12] MB
$\mathcal{Q}_{i,k}^{0}$	[15, 20] MHz	$\mathcal{Q}_{i,k}^{1}$	[25, 30] MHz

5.5 Performance Evaluation

In this section, we carry out evaluations to demonstrate the performance of the proposed learning-based algorithm for DNN inference in the end-edge-cloud orchestrated IIoT networks. We consider a smart factory with 10 IGWs deployed in this scenario, where each IGW is equipped with a Raspberry Pi 4B [46] and connected with 35 IIoT devices. We deploy an ECoS and an ECaS on the BS, and deploy a CCaS on the cloud. All these nodes are equipped with a certain amount of resources for supporting the inference task assignment and pre-trained DNN model placement, and the ratio of the noisy data is set to 30%. In the simulation, we set the performance metrics of pre-trained DNN models according to [4, 47], which are listed in Table 5.3. Here, the parameters of each pre-trained DNN model are sampled uniformly from the corresponding value interval, which follows the sampling rules described in Section III-A.

For the proposed TD3-based algorithm, we deploy a four-layer FCNN with [1800, 2400, 3000, 3500] neurons as the actor, and deploy two five-layer FCNNs with [4800, 3500, 1500, 500, 1] neurons as the critics. Here, the neurons in the output layer of the actor are activated by the $tanh$ function, and other layers' neurons of the actor and critics are activated by the $ReLU$ function. All the network parameters are trained using Adam optimizer [45]. In addition, the variance ρ of the added noise is set to 0.001, and the clip bound c is set to 0.03. Other simulation parameters of the network and learning algorithm are listed in Table 5.4. To show the efficiency of the proposed algorithm, the following benchmarks are utilized for performance comparison.

- *DDPG*: The edge controller allocates the spectrum, computing, and caching resources with the deterministic policy, where we deploy a four-layer FCNN with [1800, 2400, 3000, 3500] neurons as the actor and a four-layer FCNN with [2000, 1500, 500, 1] neurons as the critic to achieve the optimal performance.
- *Random*: In the random policy, the edge controller randomly assigns the inference tasks to IGWs or ECoS, selects an accuracy level of DNN model for each task, and allocates the multi-dimensional resources for the inference tasks of IIoT devices. All available actions are selected with an equal probability.

Fig. 5.5 Average inference accuracy and task completion ratio over inference tasks assigned to the IGWs

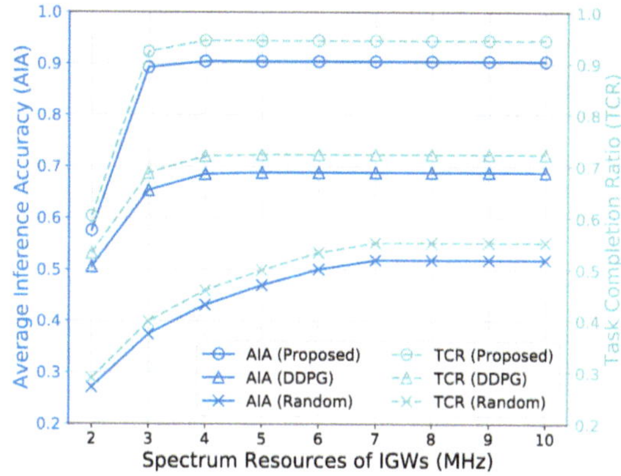

(a) Impact of the spectrum resources of IGWs

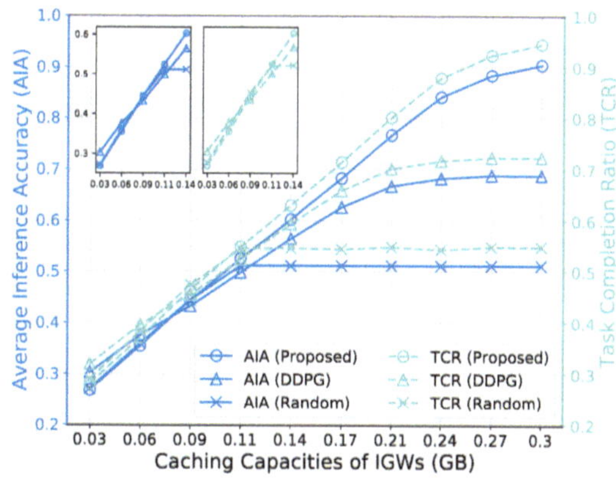

(b) Impact of the caching capacities of IGWs

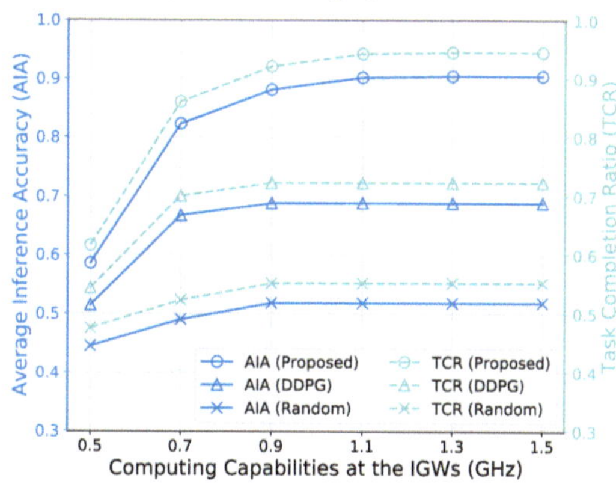

(c) Impact of the computing capabilities of IGWs

Fig. 5.6 Average inference accuracy over inference tasks assigned to the BS (i.e., ECaS and ECoS)

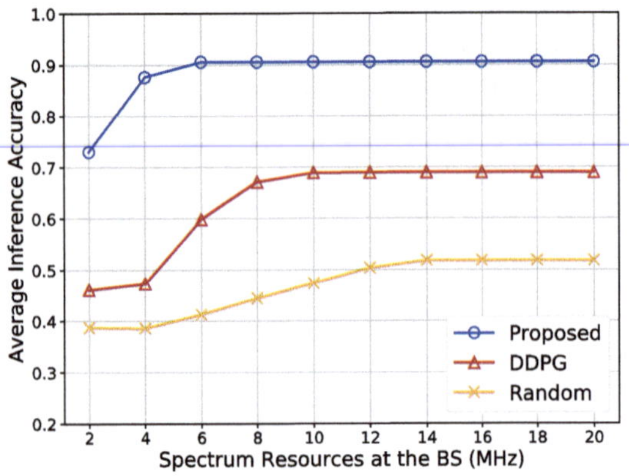

(a) Impact of the spectrum resources of BS

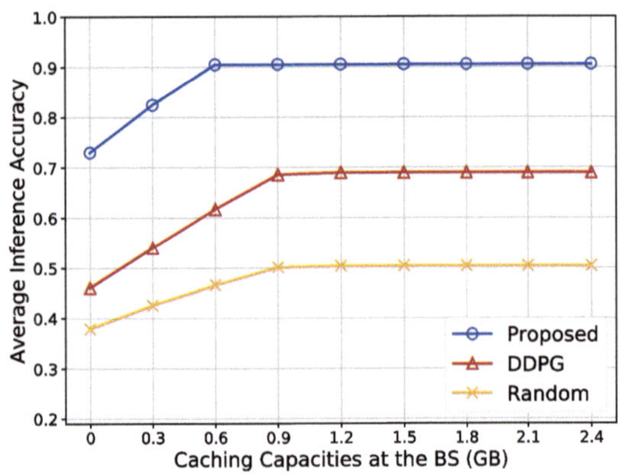

(b) Impact of the caching capacities of BS

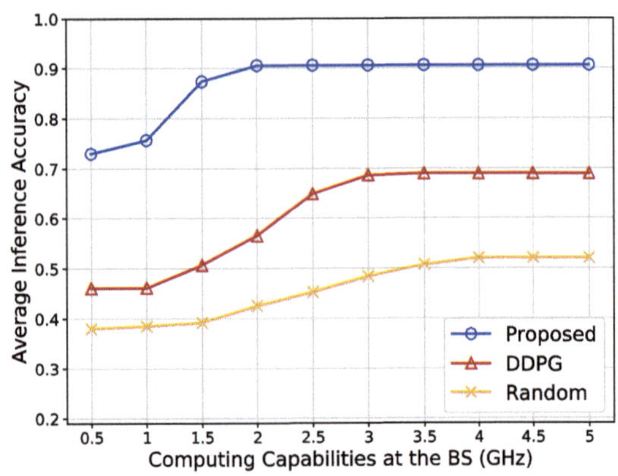

(c) Impact of the computing capabilities of BS

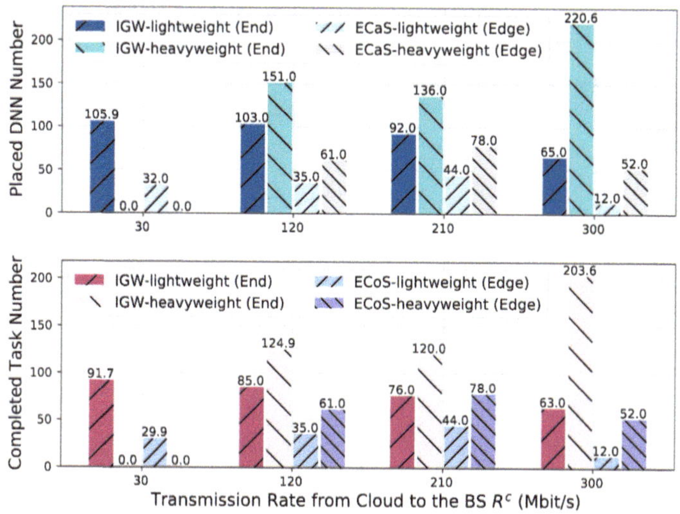

(a) The impact of the transmission rate R^c

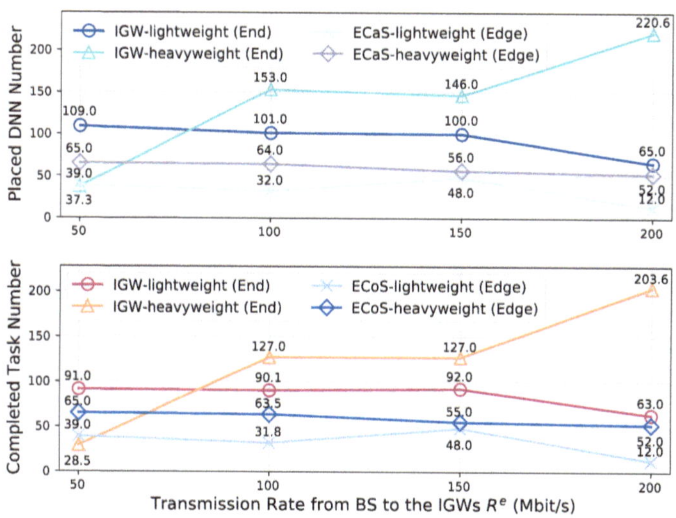

(b) The impact of the transmission rate R^e

Fig. 5.7 Numbers of completed inference tasks and placed DNN models at the IGWs and BS with respect to the transmission rates of R^c and R^e

1. *Convergence of the proposed algorithm*: As shown in Fig. 5.4, the convergence performance of the proposed TD3-based algorithm is presented with respect to training episodes. It can be seen from the red curve that the reward fluctuates in the first fifty episodes and tends to reach a stable level after parameter iterations for

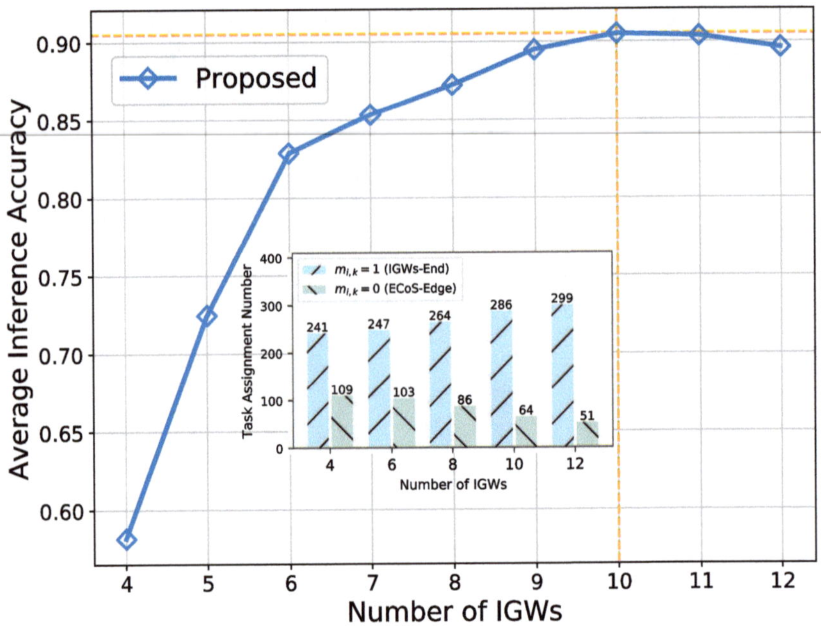

Fig. 5.8 Average inference accuracy of the proposed TD3-based algorithm with respect to the number of IGWs

Table 5.4 Simulation parameters [35, 44]

Network parameter	Value	Learning Parameter	Value
$p_{i,k}$, $p^e_{i,k}$	30, 40 dBm	*episode*	300
σ^2	− 104 dBm	*step*	50
B, B^e	10, 20 MHz	D_r	10,000
$\mathcal{C}^l$, $\mathcal{C}^e$	1.5, 5 GHz	D_b	64
$\mathcal{S}^l$, $\mathcal{S}^e$	0.3, 2.4 GB	γ	0.95
R^e, R^c	200, 300 Mbps	ϵ_a, ϵ_c	1e-4, 3e-4
$D_{i,k}$	[40, 60] KB	τ	0.01
$T^{\max}_{i,k}$	[1.5, 2] s	d	2

130 episodes. Additionally, compared with DDPG-based scheme, the proposed algorithm can not only achieve a faster convergence speed, but also obtain a higher converged value which is larger than that of the DDPG benchmark by 31.5%. This is because three improvements of the TD3 algorithm mentioned in

Section IV-B effectively mitigate possible overestimation, thereby achieving a faster learning speed and better performance than the DDPG benchmark.

2. *Impact of spectrum, caching, and computing resources at the IGWs*: We use the following evaluation metrics to measure the resource management performance of the proposed algorithm: (1) average inference accuracy; and (2) task completion ratio which is defined as the number of inference tasks completed within the delay requirements over the total number of tasks.
   As shown in Fig. 5.5, the achievable average inference accuracy and task completion ratio are presented with respect to the spectrum, caching, and computing resources at the IGWs, respectively. As can be seen from Fig. 5.5 (a-c), with the increasing of the available spectrum/caching/computing resources of IGWs, the average inference accuracy and task completion ratio achieved by the proposed TD3-based algorithm and both benchmarks increase. This is because more spectrum resources can be allocated to the IIoT devices to offload the tasks, more caching capacities can be provided for pre-trained DNN model placement, and more computing resources can be used to execute the inference tasks at IGWs. Particularly, the proposed algorithm can improve the average inference accuracy by 31.4% and 74.5%, respectively, as compared to the DDPG and Random schemes. In addition, the proposed algorithm achieves 1.31× and 1.71× task completion ratio improvement compared with DDPG and Random schemes, respectively. When the IGWs are equipped with sufficient spectrum, caching, and computing resources, inference tasks can be offloaded to the IGWs with a low latency, more pre-trained DNN models can be placed at the IGWs from ECaS or CCaS, and adequate computing resources can be utilized to accelerate DNN inference. Thus, urgent inference tasks can be executed at the end side within the strict delay requirements.
3. *Impact of spectrum, caching, and computing resources at the BS*: As shown in Fig. 5.6, we compare the achievable average inference accuracy with benchmark schemes in terms of the spectrum, caching, and computing resources of BS, respectively. Similar to the IGWs, with the increasing of the available spectrum/caching/computing resources, the average inference accuracy achieved by the TD3-based algorithm and the compared schemes also increases. The reason is that more resources can be utilized to satisfy the requirements of inference tasks that are assigned to the BS. In particular, the proposed algorithm can achieve a converged average inference accuracy faster than the benchmark schemes. Specifically, when the amount of spectrum resource is 6 MHz, the proposed algorithm achieves 1.51× and 2.20× average inference accuracy improvement compared with DDPG and Random schemes, respectively. When the caching capacity is 0.6 GB, the proposed algorithm achieves 1.48× and 1.95× average inference accuracy improvement compared with the benchmarks, respectively. When the computing capability is 2 GHz, the proposed algorithm achieves 1.60× and 2.14× average inference accuracy improvement compared with the benchmarks, respectively. The above observations indicate that the proposed algorithm can assign inference tasks and place DNN models among network nodes more reasonably, and can utilize multi-dimensional resources more efficiently, thus

faster converged average inference accuracy can be reached with a low resource demand.

4. *Impact of transmission rates of* R^c and R^e: As shown in Fig. 5.7, the distribution of inference task assignment and pre-trained DNN placement is presented in terms of the transmission rates from cloud to the BS (i.e., R^c) and from BS to the IGWs (i.e., R^e). For the upper part of Fig. 5.7a and b, the number of DNN models which are placed at the IGWs or ECaS is presented in terms of the R^c and R^e. For the below part of Fig. 5.7a and b, the number of completed tasks which are assigned to the IGWs or ECoS is presented in terms of the R^c and R^e. Several observations can be captured from Fig. 5.7. First, with the increasing of R^c and R^e, more pre-trained DNN models can be placed at the IGWs and ECaS, and more inference tasks are executed within the strict delay requirements, thus the task completion number can be improved. Second, if the network is equipped with larger transmission rate R^c, the edge controller is inclined to select heavy-weight DNN models for maximizing the average inference accuracy due to the accuracy-optimal objective. It is interesting to note that, when the R^e increases to 150 Mbit/s, the edge controller tends to execute more light-weight DNNs at the ECoS to improve the average inference accuracy. Third, if R^c and R^e are large enough, the edge controller can place the pre-trained DNN models to the IGWs with low latency, thus more inference tasks can be executed at the end side, and more heavy-weight DNN models can be selected to maximize the average inference accuracy. For example, when the R^c and R^e are set to 300 and 200 Mbit/s, the edge controller assigns 267 inference tasks to the IGWs, which is larger 4.17× than the assignment number of ECoS. Meanwhile, the edge controller selects 256 heavy-weight DNN models which is larger 3.41× than the selection number of light-weight DNN models. In particular, the utilization of the placed DNN models of IGWs achieves 93.3% for R^c = 300 Mbit/s and R^e =200 Mbit/s. The result indicates that the proposed algorithm achieves a high utilization of the placed DNN models, which effectively supports the real-time DNN inference at the end devices or the edge nodes.
5. *Impact of the number of IGWs*: As shown in Fig. 5.8, the impact of the number of IGWs on the achievable average inference accuracy of the proposed TD3-based algorithm is presented, where the number of IIoT devices connected to each IGW is the same. It can be seen that the average inference accuracy increases with the number of IGWs when it is less than 10, which is because more resources can be allocated to the inference tasks. In addition, the average inference accuracy improvement introduced by each IGW diminishes with the increasing number of IGWs, and a decline trend appears after the number of IGWs exceeds 10. This is because more caching and computing resources are deployed at the end side with the increasing number of IGWs. In such a case, the edge controller is inclined to make decisions to assign inference tasks to the IGWs. The embedded figure shows the distribution of task assignment decisions with respect to the number of IGWs. However, to perform DNN inference at the end side, more pre-trained DNN models need to be placed to the IGWs from the CCaS or ECaS, thus urgent

inference tasks cannot be completed within the strict delay requirements due to the large transmission delay of DNN models.

5.6 Summary

In this chapter, we have studied the joint TADP problem for DNN inference in the end-edge-cloud orchestrated IIoT networks with multi-dimensional resource constraints. A learning-based algorithm has been proposed to make resource allocation decisions in real time, such that providing DNN inference services for intelligent IIoT applications. In resource-limited IIoT networks, the proposed algorithm can learn the optimal DNN model placement scheme, and select appropriate DNN models for inference tasks via efficient resource utilization, thereby effectively improving the average accuracy of inference tasks while satisfying the strict delay requirements. For the future work, we will investigate the distributed DNN training problem in IIoT networks.

References

1. N. Kato, B. Mao, F. Tang, Y. Kawamoto, J. Liu, Ten challenges in advancing machine learning technologies toward 6G. IEEE Wireless Commun. **27**(3), 96–103 (2020)
2. K. Zhang, Y. Zhu, S. Maharjan, Y. Zhang, Edge intelligence and blockchain empowered 5G beyond for the industrial Internet of Things. IEEE Netw. **33**(5), 12–19 (2019)
3. E. Sisinni, A. Saifullah, S. Han, U. Jennehag, M. Gidlund, Industrial Internet of things: challenges, opportunities, and directions. IEEE Trans. Ind. Informat. **14**(11), 4724–4734 (2018)
4. W. Zhang, D. Yang, H. Wang, Data-driven methods for predictive maintenance of industrial equipment: a survey. IEEE Syst. J. **13**(3), 2213–2227 (2019)
5. W. Zhang, D. Yang, Y. Xu, X. Huang, J. Zhang, M. Gidlund, DeepHealth: a self-attention based method for instant intelligent predictive maintenance in industrial Internet of things. IEEE Trans. Ind. Informat. **17**(8), 5461–5473 (2021)
6. C. Xu, K. Ma, Y. Xu, Y. Xu, Y. Fang, Optimal power scheduling for uplink transmissions in SIC-based industrial wireless networks with guaranteed real-time performance. IEEE Trans. Veh. Technol. **67**(4), 3216–3228 (2018)
7. Y. Su, W. Fan, L. Gao, L. Qiao, Y. Liu, F. Wu, Joint DNN partition and resource allocation optimization for energy-constrained hierarchical edge-cloud systems. IEEE Trans. Veh. Technol. **72**(3), 3930–3944 (2022)
8. P. Ren, L. Liu, X. Qiao, J. Chen, Distributed edge system orchestration for web-based mobile augmented reality services. IEEE Trans. Serv. Comput. **16**(3), 1778–1792 (2022)
9. L. Wang, X. Ren, C. Zhao, F. Zhao, S. Yang, MPDM: a multi-paradigm deployment model for large-scale edge-cloud intelligence. IEEE Internet Things J. **10**(10), 8773–8785 (2022)
10. G. Zheng, M. Wen, Z. Ning, and Z. Ding, Computation-aware offloading for DNN inference tasks in semantic communication assisted MEC systems. IEEE Trans. Wireless Commun. (2025)
11. L. Zeng, C. Yang, P. Huang, Z. Zhou, S. Yu, X. Chen, GNN at the edge: cost-efficient graph neural network processing over distributed edge servers. IEEE J. Sel. Areas Commun. **41**(3), 720–739 (2022)

12. C. Wang, Z. Yuan, P. Zhou, Z. Xu, R. Li, D.O. Wu, The security and privacy of mobile-edge computing: an artificial intelligence perspective. IEEE Internet Things J. **10**(24), 22008–22032 (2023)
13. Y.-T. Yang, H.-Y. Wei, A coalition formation approach for privacy and energy-aware split deep learning inference in edge camera network. IEEE Trans. Netw. Serv. Manag. **20**(3), 3673–3685 (2022)
14. L. Zeng, X. Chen, P. Huang, K. Luo, X. Zhang, Z. Zhou, Serving graph neural networks with distributed fog servers for smart IoT services. IEEE/ACM Trans. Netw. **32**(1), 550–565 (2023)
15. V. Sze, Y. Chen, T. Yang, J. Emer, Efficient processing of deep neural networks: a tutorial and survey. Proc. IEEE **105**(12), 2295–2329 (2017)
16. E. Li, L. Zeng, Z. Zhou, X. Chen, Edge AI: on-demand accelerating deep neural network inference via edge computing. IEEE Trans. Wireless Commun. **19**(1), 447–457 (2020)
17. Y. Li, Z. Han, Q. Zhang, Z. Li, H. Tan, *"Automating Cloud Deployment for Deep Learning Inference of Real-Time Online Services, in Proc* (IEEE INFOCOM, Las Vegas, United States, 2020), pp. 1668–1677
18. N. Zhang, R. Wu, S. Yuan, C. Yuan, D. Chen, RAV: relay aided vectorized secure transmission in physical layer security for Internet of things under active attacks. IEEE Internet of Things J. **6**(5), 8496–8506 (2019)
19. T. Taleb, K. Samdanis, B. Mada, H. Flinck, S. Dutta, D. Sabella, On multi-access edge computing: a survey of the emerging 5G network edge cloud architecture and orchestration. IEEE Commun. Surveys Tuts. **19**(3), 1657–1681 (2017)
20. J. Wang, L. Zhao, J. Liu, N. Kato, Smart resource allocation for mobile edge computing: a deep reinforcement learning approach. IEEE Trans. Emerg. Topics Comput. **9**(3), 1529–1541 (2021)
21. W. Zhuang, Q. Ye, F. Lyu, N. Cheng, J. Ren, SDN/NFV-empowered future IoV with enhanced communication, computing, and caching. Proc. IEEE **108**(2), 274–291 (2020)
22. N. Luong, D. Hoang, S. Gong, D. Niyato, P. Wang, Y. Liang, D. Kim, Applications of deep reinforcement learning in communications and networking: a survey. IEEE Commun. Surveys Tuts. **21**(4), 3133–3174 (2019)
23. K. Li, W. Ni, E. Tovar, A. Jamalipour, *"Deep Q-Learning Based Resource Management in UAV-Assisted Wireless Powered IoT Networks, in Proc* (IEEE ICC, Dublin, Ireland, 2020), pp. 1–6
24. N. Cheng, F. Lyu, W. Quan, C. Zhou, H. He, W. Shi, X. Shen, Space/aerial-assisted computing offloading for IoT applications: a learning-based approach. IEEE J. Sel. Areas Commun. **37**(5), 1117–1129 (2019)
25. W. Wu, N. Chen, C. Zhou, M. Li, X. Shen, W. Zhuang, X. Li, Dynamic RAN slicing for service-oriented vehicular networks via constrained learning. IEEE J. Sel. Areas Commun. **39**(7), 2076–2089 (2021)
26. Y. Zhou, Z. Fadlullah, B. Mao, N. Kato, A deep-learning-based radio resource assignment technique for 5G ultra dense networks. IEEE Netw. **32**(6), 28–34 (2018)
27. Y. He, N. Zhao, H. Yin, Integrated networking, caching, and computing for connected vehicles: a deep reinforcement learning approach. IEEE Trans. Veh. Technol. **67**(1), 44–55 (2018)
28. W. Wu, N. Cheng, N. Zhang, P. Yang, W. Zhuang, X. Shen, Fast mmwave beam alignment via correlated bandit learning. IEEE Trans. Wireless Commun. **18**(12), 5894–5908 (2019)
29. W. Zhang, D. Yang, H. Peng, W. Wu, W. Quan, H. Zhang, X. Shen, *"Deep Reinforcement Learning Based Resource Management for DNN Inference in IIoT, in Proc* (IEEE GLOBECOM, Taipei, Taiwan, 2020)
30. A. Eshratifar, M. Abrishami, M. Pedram, JointDNN: an efficient training and inference engine for intelligent mobile cloud computing services. IEEE Trans. Mobile Comput. **20**(2), 565–576 (2021)
31. Y. Huang, F. Wang, F. Wang, J. Liu, *"Deepar: A Hybrid Device-Edge-Cloud Execution Framework for Mobile Deep Learning Applications, in Proc* (IEEE INFOCOM, Paris, France, 2019), pp. 892–897

32. W. Sun, J. Liu, Y. Yue, AI-enhanced offloading in edge computing: when machine learning meets industrial IoT. IEEE Netw. **33**(5), 68–74 (2019)
33. B. Yang, X. Cao, X. Li, Q. Zhang, L. Qian, Mobile edge computing based hierarchical machine learning tasks distribution for IIoT. IEEE Internet Things J. **7**(3), 2169–2180 (2020)
34. Y. You, Z. Zhang, C. Hsieh, J. Demmel, K. Keutzer, Fast deep neural network training on distributed systems and cloud TPUs. IEEE Trans. Parallel Distrib. Sys. **30**(11), 2449–2462 (2019)
35. H. Peng, X. Shen, Deep reinforcement learning based resource management for multi-access edge computing in vehicular networks. IEEE Trans. Netw. Sci. Eng. **7**(4), 2416–2428 (2020)
36. X. Shen, J. Gao, W. Wu, K. Lyu, M. Li, W. Zhuang, X. Li, J. Rao, AI-assisted network-slicing based next-generation wireless networks. IEEE Open J. Veh. Technol. **1**(1), 45–66 (2020)
37. W. Wu, P. Yang, W. Zhang, C. Zhou, X. Shen, Accuracy-guaranteed collaborative DNN inference in industrial IoT via deep reinforcement learning. IEEE Trans. Ind. Informat. **17**(7), 4988–4998 (2021)
38. Y. Liu, H. Yu, S. Xie, Y. Zhang, Deep reinforcement learning for offloading and resource allocation in vehicle edge computing and networks. IEEE Trans. Veh. Technol. **68**(11), 11 158–11 168 (2019)
39. N. Zhao, Y. Liang, D. Niyato, Y. Pei, M. Wu, Y. Jiang, Deep reinforcement learning for user association and resource allocation in heterogeneous cellular networks. IEEE Trans. Wireless Commun. **18**(11), 5141–5152 (2019)
40. J. Chen, Z. Wei, S. Li, B. Cao, Artificial intelligence aided joint bit rate selection and radio resource allocation for adaptive video streaming over F-RANs. IEEE Wireless Commun. **27**(2), 36–43 (2020)
41. Z. Zhang, H. Chen, M. Hua, C. Li, Y. Huang, L. Yang, Double coded caching in ultra dense networks: caching and multicast scheduling via deep reinforcement learning. IEEE Trans. Commun. **68**(2), 1071–1086 (2019)
42. D. Yang, Y. Xu, H. Wang, T. Zheng, H. Zhang, H. Zhang, M. Gidlund, Assignment of segmented slots enabling reliable real-time transmission in industrial wireless sensor networks. IEEE Trans. Ind. Electron. **62**(6), 3966–3977 (2015)
43. T. Lillicrap, J. Hunt, A. Pritzel, N. Heess, T. Erez, Y. Tassa, D. Silver, D. Wierstra, *"Continuous Control with Deep Reinforcement Learning, in Proc* (ICLR, San Juan, Puerto Rico, 2016)
44. S. Fujimoto, H. Van, D. Meger, Addressing function approximation error in actor-critic methods, arXiv preprint arXiv:1802.09477 (2018)
45. D. Kingma, J. Ba, Adam: a method for stochastic optimization, arXiv preprint arXiv:1412.6980 (2014)
46. Raspberry Pi 4 Model B. [Online]. Available: https://www.raspberrypi.org/documentation/hardware/raspberrypi/bcm2711/rpi_DATA_2711_1p0_preliminary.pdf. Accessed on: 21 Jun 2019
47. J. Luo, J. Wu, W. Lin, *"Thinet: A Filter Level Pruning Method for Deep Neural Network Compression, in Proc* (ICCV, Venice, Italy, 2017), pp. 5058–5066
48. W. Wu, N. Zhang, N. Cheng, Y. Tang, K. Aldubaikhy, X. Shen, Beef up mmWave dense cellular networks with D2D-assisted cooperative edge caching. IEEE Trans. Veh. Technol. **68**(4), 3890–3904 (2019)
49. W. Quan, N. Cheng, M. Qin, H. Zhang, H. Chan, X. Shen, Adaptive transmission control for software defined vehicular networks. IEEE Wireless Commun. Lett. **8**(3), 653–656 (2018)

Chapter 6
Conclusions

6.1 Summary of the Research

This book provides a comprehensive exploration of DNN applications and resource scheduling strategies tailored for IIoT. Focusing on the entire lifecycle of DNN models, we have proposed innovative network architectures and resource scheduling algorithms to optimize both the training and inference phases of DNNs within IIoT environments. These advancements aim to enhance overall system performance and resource utilization efficiency.

In Chap. 2, we propose a network architecture that integrates communication and computation, named Com^{2}Net, which uses quantum genetic algorithms to achieve intelligent scheduling and resource allocation of computing tasks across space-air-ground, end-edge-cloud, and multi-data center domains. Com^{2}Net not only improves the efficiency of completing computing tasks, but also significantly improves resource utilization, providing new ideas for the design of future network architectures. This research emphasizes the importance of deep integration of communication and computation, and lays the foundation for building efficient and flexible future networks.

Concentrating on the resource allocation problem during the FL process in distributed IIoT, Chap. 3 introduces a three-layer collaborative FL architecture to support the training of DNN models. In addition, an optimization scheme based on multi-agent DRL is proposed. This scheme effectively minimizes the evaluation loss of federated learning by jointly optimizing device selection and computing and spectrum resource allocation, while satisfying strict latency and long-term energy consumption requirements.

In response to the requirements of time-sensitive IIoT, Chap. 4 proposes a deterministic FL framework, named DetFed, which adopts a two-layer aggregation mechanism and integrates wireless and wired integrated TSN to achieve deterministic transmission guarantee of FL parameters. In addition, a DRL-based resource scheduling

© The Author(s), under exclusive license to Springer Nature Switzerland AG 2026
W. Zhang et al., *Intelligent Resource Scheduling in End-Edge-Cloud Networks*,
Wireless Networks, https://doi.org/10.1007/978-3-032-07667-0_6

algorithm has been proposed to make the optimal temporal resource allocation decisions for facilitating efficient DetFed, which aggregates more device-side models for obtaining an accurate global model.

In Chap. 5, we focus on the real-time DNN inference phase in the IIoT and propose a resource management scheme based on DRL. This solution flexibly coordinates the allocation of inference tasks and the placement of DNN models through the end-edge-cloud collaborative architecture, aiming to maximize the average accuracy of inference tasks while satisfying strict latency requirements.

6.2 Discussion of Future Directions

There are many potential future directions for resource scheduling in end-edge-cloud networks. Here, we elaborate on four promising directions, as follows.

The first is blockchain-enabled computing resource transaction. With the development of blockchain technologies, dispersive and heterogeneous computing resources can be securely trade on a public platform. Users can publish their requests for computing resources on the platform, and suppliers can sell or rent their available resources, thus realizing the schedulable, exchangeable, and on-demand resource supply. However, when misleading information is published, it will result in an imbalance between supply and demand, which is not conducive to efficient platform management. How to construct a trustworthy and transparent computing transaction platform is an interesting topic [1, 2].

The second is AI-assisted network optimization. In recent years, research works on network resource optimization to support the construction of large AI models have received widespread attention, and many state-of-the-art schemes have been proposed to improve resource utilization efficiency. Similarly, large AI models, with their powerful data processing and learning capabilities, are expected to become a new driver of resource optimization in future networks, which have the potential to achieve dynamic resource allocation and scheduling while adapting to the changing requirements of computing tasks. In this context, how to deeply embed large AI models into the network architecture rather than as an external enhancement is worthy of further research. For example, traffic scheduling in 6G networks can be accomplished by deploying AI agents in switches and routers, thereby facilitating the efficient transmission of multimodal data. Therefore, the construction and application of large AI models in future networks will be more flexible and more effective [3, 4].

The third is lightweight model deployment optimization driven by edge intelligence. On the one hand, due to the limited computing resources of edge devices, the compression and pruning technology of DNN models requires study to reduce the model size and computing workload, which is convenient for efficient deployment on edge devices. On the other hand, the collaborative inference mechanism between edge devices can be explored to improve the overall inference efficiency and accuracy through task segmentation and result fusion, while reducing the dependence on the cloud [5–7].

The fourth is green collaborative computing. In the construction of large AI models, energy consumption has become an increasingly serious issue. Recently, to realize energy-efficient training, deployment, and inference for large AI models, green collaborative computing has received wide attention. Through distributed computing, model parallelism, and resource sharing, the unnecessary waste of computing resources can be greatly reduced. Hence, how to develop the corresponding solutions to support green collaborative computing is an important research issue. Moreover, multi-model joint training and cross-device collaborative inference are worth discussing to reduce resource consumption while ensuring model performance [8–10].

References

1. Y. Zuo, Exploring the synergy: AI enhancing blockchain, blockchain empowering AI, and their convergence across IoT applications and beyond. IEEE Internet Things J. **12**(6), 6171–6195 (2025)
2. E. Moore, A. Imteaj, M. Hossain, S. Rezapour, M. Amini, Blockchain-empowered cyber-secure federated learning for trustworthy edge computing. IEEE Trans. Artif. Intell. 1–1 (2025)
3. W. Zhang, T. Sun, D. Yang, T.H. Luan, H. Zhang, Dynamic resource scheduling for deterministic communication, computation, and control integration in industrial cyber-physical systems. IEEE Trans. Cogn. Commun. Netw. 1–1 (2025)
4. J. Ren, W. Zhang, H. Wang, D. Yang, S. Wang, H. Zhang, and S. Cui, Toward deterministic wide-area networks via deadline-aware routing and scheduling. IEEE Trans. Netw. 1–17 (2025)
5. M. Zhang, X. Shen, J. Cao, Z. Cui, S. Jiang, EdgeShard: Efficient LLM inference via collaborative edge computing. IEEE Internet Things J. **12**(10), 13 119–13 131 (2025)
6. G. Qu, Q. Chen, W. Wei, Z. Lin, X. Chen, K. Huang, Mobile edge intelligence for large language models: a contemporary survey. IEEE Commun. Surv. Tutor. 1–1 (2025)
7. Z. Nan, Y. Han, J. Yan, S. Zhou, Z. Niu, Robust task offloading and resource allocation under imperfect computing capacity information in edge intelligence systems. IEEE Trans. Mob. Comput. 1–1 (2025)
8. X. Ju, H. Liu, S. Gong, C. Xing, N. Zhao, D. Niyato, A framework for energy efficiency optimization in IRS-Aided hybrid MU-MIMO systems. IEEE J. Sel. Areas Commun. **43**(3), 883–898 (2025)
9. Y. Wang, C. Yang, S. Lan, L. Zhu, Y. Zhang, End-edge-cloud collaborative computing for deep learning: a comprehensive survey. IEEE Commun. Surv. Tutor. **26**(6), 2647–2683 (2024)
10. X. He, H. Xu, X. Xu, Y. Chen, Z. Wang, An efficient algorithm for microservice placement in cloud-edge collaborative computing environment. IEEE Trans. Serv. Comput. **17**(5), 1983–1997 (2024)

Index

© The Editor(s) (if applicable) and The Author(s), under exclusive license to Springer Nature Switzerland AG 2026

W. Zhang et al., *Intelligent Resource Scheduling in End-Edge-Cloud Networks*, Wireless Networks, https://doi.org/10.1007/978-3-032-07667-0

MIX
Papier aus verantwortungsvollen Quellen
Paper from responsible sources
FSC® C105338

If you have any concerns about our products,
you can contact us on
ProductSafety@springernature.com

In case Publisher is established outside the EU,
the EU authorized representative is:
Springer Nature Customer Service Center GmbH
Europaplatz 3, 69115 Heidelberg, Germany

Printed by Libri Plureos GmbH
in Hamburg, Germany